Dark Light

ALSO BY LINDA SIMON

Genuine Reality: A Life of William James

Dark Light

Electricity and Anxiety from the Telegraph to the X-Ray

Linda Simon

A Harvest Book • Harcourt, Inc.
Orlando Austin New York San Diego Toronto London

www.HarcourtBooks.com

Library of Congress Cataloging-in-Publication Data
Simon, Linda, 1946–
Dark light: electricity and anxiety from the telegraph to the X-ray/
Linda Simon.—1st ed.
p. cm.
Includes bibliographical references and index.
1. Electrification—Public opinion—History—19th century.
2. Electrification—Public opinion—History—20th century.
3. Electricity—Psychological aspects—History—19th century.
4. Electricity—Social aspects—History—19th century. 5. Fear. I. Title.
TK17.S56 2004
303.48'3—dc22 2003019994
ISBN 0-15-100586-9
ISBN-13: 978-0156-03244-5 (pbk.)
ISBN-10: 0-15-603244-9 (pbk.)

Text set in Bulmer MT
Designed by Cathy Riggs

Printed in the United States of America

First Harvest edition 2005
A C E G I K J H F D B

for Eva Hindus

and in memory of Milton Hindus

Contents

PART III: ELECTROSTRIKES

Dark Light

INTRODUCTION

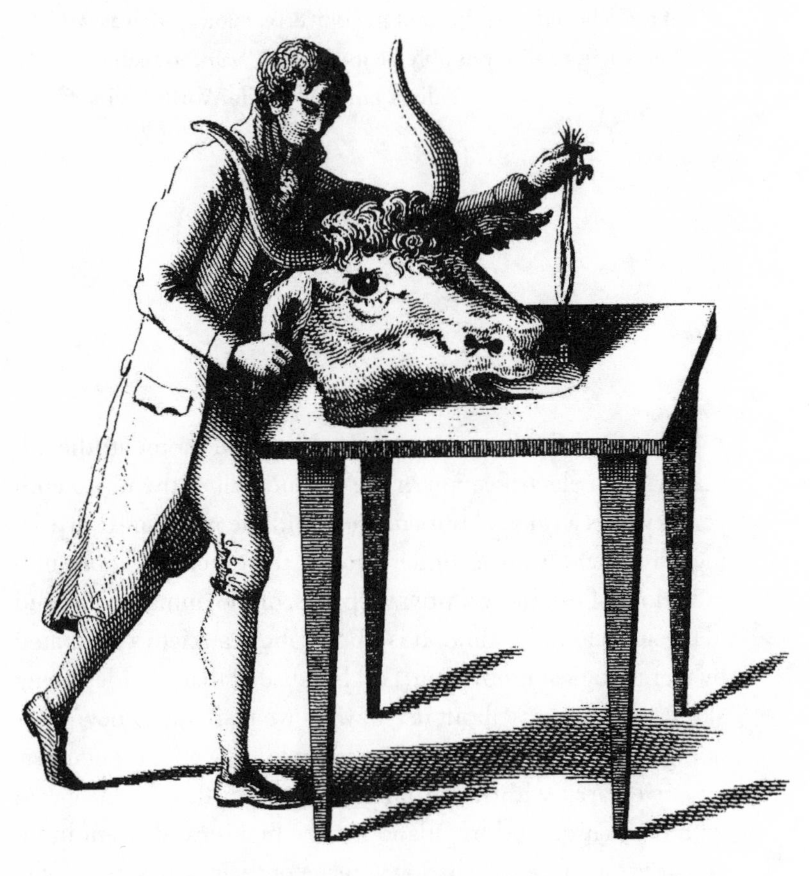

Giovanni Aldini electrifying the head of an animal. Such demonstrations led audiences to believe that electricity animated living matter.

The Bakken Library, Minneapolis

If there be a divine Spirit of the universe, nature, such as we know her, cannot possibly be its *ultimate word* to man.

William James, "Is Life Worth Living?"

This book is about a particular historical moment: the advent of electrification in the second half of the nineteenth century. It is a book about energies, and the many and surprising ways that term was understood at the time; and about illumination, of public and private spaces, of the human body, and of the spirit and the mind. It is a book about anxieties generated by technological innovation, and because of that, besides being about the past, it is about us, now. As we respond to new technologies—human cloning, for example, or genetic engineering—we carry with us an inheritance from those who gazed with fascination and trepidation at the first incandescent bulb, and at the astounding shadowy image of their bones, made visible by an inexplicable ray. *Dark Light* offers us a way to reflect upon our responses, to illuminate who we once were, and to imagine who we might become.

I became interested in electricity, oddly enough, through my work on William James. Philosopher, psychologist, and ardent psychical researcher, James wanted to understand what he called the "wild facts" in the universe—apparitions, occult powers, telepathy—phenomena that suggested the existence of mysterious energies. A world that did not resonate with these invisible forces, he believed, would be dead, cold, and lusterless. Because many of James's colleagues in psychical research were physicists whose professional work focused on electricity, and because the language describing these wild facts borrowed significantly from the language used to describe electrical phenomena, I wondered how—or if—early electrification was connected to the coincident intensification of psychical research. That question led me to histories of electrification, where I made two unexpected discoveries. First, that although electricity surely generated excitement, and although electrical companies worked hard to gain a domestic market for the power, its use spread surprisingly slowly, suggesting consumer resistance. And second, that the application of electricity as a medical therapy spread rapidly, embraced with enthusiasm by physicians and patients. This apparent contradiction, and the questions it raised, motivated me to write this book.

Dark Light looks at a time when electricity was a force stronger in the imagination than in reality, when there were few lightbulbs, telephones, or electric trams in everyday use; when people learned of the potential of electricity from the spectacular demonstrations they saw at fairs and from eager articles in magazines and daily newspapers. Readers were invited to imagine themselves in control of a benign power—a power that would protect them. Electricity's first use in homes was in batteries for burglar and fire alarms; electric lights, readers learned, would not ignite the curtains in children's bedrooms. Of course, the power would enhance comfort, too. Door openers, shoe polishers, and even a miniature train track installed around a dinner table to distribute plates of food would make everyday tasks easier. Elevators, carpet sweepers, fans, coffeepots, and sewing machines would transform housekeeping. But beyond this image of life made easy by a multiplicity of electrical gadgets, people were told that as wielders of electricity, they themselves would be transformed for the better. Once women controlled electrical power in their homes, once they freed themselves from mundane chores, they would evolve into man's intellectual equal. When electricity speeded up the pace of life, human evolution would proceed faster.

With such euphoric predictions, one would suppose that the public hardly could wait for electrical power in their homes; instead, more than thirty years after Thomas Edison invented the incandescent bulb in 1879 and soon afterward installed a lighting system in a business section of lower Manhattan, barely 10 percent of American homes were wired. Even after the First World War, that percentage rose only to 20 percent. In France, between 1889 and 1900, the number of people using gas doubled. As late as 1923, contractors' advertisements still were trying to convince an apparently reluctant American public of the benefits of electricity. "Electric light is safe," these advertisements proclaimed. Electricity would make a home healthier, more comfortable, more beautiful, more efficient, and, as advertisements repeatedly insisted, happier. "Electric light completes the happy home circle." "Electricity makes everybody smile."[1]

Since the lightbulb seems like an incontestably good idea, and since we ourselves would be horrified at giving up electrification, I wondered why the nineteenth-century public needed so much persuasion to try electricity. Some reasons were not hard to discover. Alongside articles extolling electricity were others that reflected fear, distrust of the hyperbolic claims of electrical companies, and worry about the possible physical consequences of a new, untested technology. Accidental electrocutions made front-page news, as did terrifying explosions caused by poorly insulated wires laid alongside gas mains. Articles warned about possibly malignant effects of electricity on the human body: blindness, for example, from reading by incandescent light. Electricity, moreover, was unreliable: in cities that had installed public lighting, bulbs and wires often failed, and repairs were slow and difficult. Few trained electricians were available to support the new systems. Most significantly, harnessing electricity seemed an act of transgression against the power of nature. Electricity, after all, was nature's most devastating force; lightning was electrical, and it seemed foolhardy to invite that force into one's home.

In the context of this resistance to electrification, the widespread appeal of electrotherapy as a mainstream medical treatment seemed even more puzzling. I knew that William James and many members of his family had undergone electrical therapy: James for heart trouble, his brother Henry for digestive problems, their sister Alice for depression. Now I discovered that electrotherapy was not just one of the medical fads that the Jameses were prone to embrace but a treatment endorsed by traditional medicine throughout America and Europe. Physicians routinely included a battery set among their office equipment, and major hospitals boasted an electrical unit, headed by a physician who took the title of Electrician. This form of electrotherapy was not the electroconvulsive treatment that came into use in the 1950s for severe and intractable mental illness, but included a variety of massages, baths, showers, and infusions. Sometimes gentle and benign, sometimes painful and punitive, electrotherapy was the treatment of choice for scores of common ailments, and especially for the newly diagnosed illness neurasthenia, or nerve weakness. Electricity, physicians believed, provided nourishment for depleted nerves.

I began research, then, with many questions about what the public—those who could afford to wire their houses and pay for electrotherapy treatments—believed about electricity in the second half of the nineteenth century.[2] What did they understand about it as a physical force? What did they believe about the relationship between electrical energy and human energy? Where did their beliefs come from? What intellectual and cultural legacy shaped their responses to electrification?

Resistance to a new technology seemed, on the surface, understandable: change is often difficult and threatening, whatever that change is. Steam power and rail travel, for example, had incited fear and apprehension earlier in the century, and continued to be seen by some people as noisy, filthy intrusions. Once America had been bucolic and pastoral; now, with trains roaring across the land and steam engines bellowing, the nation was transformed for the worse.

Since electrical technology resulted from scientific research, I wondered about the public's attitudes toward science. What did people believe about the purposes of scientific discovery and the enterprise of uncovering nature's secrets? Did they imagine scientists to be arrogant Fausts or malevolent Frankensteins, treading where mortals should not venture? Or were they benefactors bringing good news of the universe to the rest of humankind? What were the proper consequences of investigation? Metaphysical insight? Ease and wealth? Cures for disease and despair? Answers to these questions complicated what appeared simply to be apprehension about the new.

To consider the public's acceptance of electrotherapy, I needed to know about mid-nineteenth-century medicine and the kinds of treatments available. If I had gone to my physician complaining of insomnia and fatigue, what would persuade me to sit in his office with electrodes on my back and forehead for half an hour, and then to return for a series of infusions week after week after week? How would that battery-produced power affect my sense of personal empowerment? If my symptoms reflected psychological and emotional distress—"morbid fears," for example, which were a popular complaint of neurasthenics—how would electricity serve me? How would I understand the causes of my distress?

These questions led me to examine beliefs—about vitalism, spiritualism, and science—that provide a crucial context for the history of electrification. Vitalism, which held that electricity was the source of life itself, justified the conviction that applying electricity to the human body could strengthen and energize; at the same time, vitalism contributed to a fear of artificial electricity, since the force that energized this new technology was, in the popular mind, the same force that coursed through one's nerves. While people were willing to submit to the invasion of electricity into their bodies, with their own permission and administered by a physician whom they trusted, they were suspicious about allowing electricity to flow into their homes, where the force could stealthily invade their body in ways that they might not be able to control. It might even kill them.

Spiritualism intensified the beliefs of vitalism, asserting that one's soul, will, and spirit were electrical in nature: the spirit of the dead existed as disembodied energies; a person with a strong electrical force manifested a powerful will that could manipulate others, infiltrate their minds, and overwhelm them. The association of electricity with the occult and with death inspired fascination, but also created an aura of fear.

Science, too, inspired fascination, offering stunning displays of electricity; but scientists' projects often seemed arrogant and foolhardy: to tamper with the electrical force risked causing an unbalance in nature; to tamper with nature at all risked punishment and retribution. Were accidental electrocutions really accidental? Or were they nature's way of striking down transgressors? Scientists comprised a new class of experts whose discoveries about nature were publicized as having vast potential to change the world morally, socially, politically, and materially. Such changes in themselves generated anxiety, but one risk seemed especially terrifying: the possibility of a universe consisting of matter, and nothing more; a universe without a guiding intelligence, a universe without God. Passionately held and often contradictory assumptions about empowerment and authority; about the wisdom of manipulating nature; about the invisible, supernormal, and magical; about the consequences of a materialist universe; and about the nature of life itself—all these beliefs shaped the public's response to electrification.

Dark Light is bracketed by Samuel Morse's development of a telegraph, the first electrical instrument offered for public, not only business, use, and the first instrument that reveals a pattern of response repeated with subsequent innovations. The telegraph pressed the public to consider what electricity could do *for* them and what it might do *to* them. I end my story with Wilhelm Roentgen's discovery of unknown rays with the inexplicable power to penetrate clothing and skin. With the X-ray, which burst into the culture at the same time as psychoanalysis, popular fascination with the hidden mysteries of the self lessened the fears—of invasion, loss of privacy, and erosion of authority—with which people had responded to previous innovations.

I've told this story about our culture's response to electricity in three parts: the first, "Wonders," provides a chronological history of the public's encounters with electricity in the form of the telegraph, phonograph, lighting, and electrotherapy. Part Two, "Cravings of the Heart," considers prevalent assumptions about the body, in sickness and health, and the body's connection to electricity; beliefs about the mind, soul, and religion; and tensions generated by scientists' apparently unquenchable desire for knowledge of the natural world. Part Three, "Electrostrikes," focuses on manifestations of cultural anxiety generated by the first criminal execution, the hyperbolic claims incited by exhibitions of electricity at the Chicago World's Fair, and the mysterious X-ray.

During the fifty years between Samuel Morse's development of the telegraph and Wilhelm Roentgen's discovery of the X-ray, the public's most intense relationship with electricity came from publicity about and demonstrations of electric lights. Illumination is the technology central to my story, and Thomas Edison—scientist, inventor, ruthless businessman, and, in the popular imagination, magician—is a central protagonist. Physician George Beard is the other central character, whose renown rests on his identification of neurasthenia, his term for the depression and anxiety brought about, he believed, by the psychological stresses of an electrified world. Beard's work on the somatic consequences of stress influenced the nascent field of psychology. Beard, coincidentally with Freud, proposed an etiology for neurasthenia that included sexual disturbances and repression. Freud knew Beard's work, and in psychotherapeutic techniques and theory, we can see that work extended.

Fear of electricity may seem quaint to us. We love our lightbulbs, and we would not happily give up our computers, dishwashers, and microwave ovens. Yet truth be told, we, too, worry about invisible emanations—from power lines or from the cell phones that we carry with us everywhere. New technologies unsettle us. We press pharmaceutical companies to find cures for everything that ails us, and even our pets, but we worry about eating genetically modified corn. Some of us are eager to swallow whatever will keep us thin, youthful, and sexually vital, but we would prefer not to swallow irradiated beef. We celebrate the mapping of the human genome, but we worry about living in a universe in which we are nothing more than our DNA. We read the day's astrological forecast, occasionally knock on wood, and harbor one or another belief about the way nature works that we would not share with our colleagues at the water cooler or the faculty club. Many of us have inherited from our nineteenth-century forebears some cherished assumptions: that nature will and must keep its secrets, that humans should not transgress, that nature will enact punishment against those who try to control it or probe too closely. Humans should not play God, we warn those who advocate manipulating bodies and minds in ways that we believe are unnatural. We worry, for reasons that are not new.

Background, Briefly

VITALISM

Life is a problem, and so is death. A fly buzzes: clearly, it is alive. But swat it, and it falls to the ground. Whatever generated its movement was obliterated in that swat. Whatever obliterated its life generates a question: What animated this insect? The fly is, from minute observation, the same physical entity before the assault as after, composed of the same chemicals, arranged in the same physical structure. What gave it the ability to move, to grow, and to reproduce itself? Are those three manifestations caused by the same force? Are they the result of internal mechanical or chemical processes? Or is there a different agent, a "vital stimulus," analogous to light and heat, that permeates living bodies or acts upon them?

These questions, inherited from generations of natural philosophers, remained unresolved and emotionally debated throughout the nineteenth century. Some believed that nature could be explained by physical and chemical models, derived from an inquiry into matter itself. Others believed that theories based on the interactions of atoms failed to consider two significant phenomena: the nature of consciousness and the relationship between movement and will. What process or substance allowed will to produce motion? What were thought, feeling, emotion, and self-awareness? What and where was the human soul? Such questions, of course, were bound up in metaphysics and religion. Identifying an animating force external to living things might provide evidence for an afterlife, immortality, and, above all, an Intelligence, a Creative Spirit: God. The quest for such a force has a long history and an enduring appeal.

The universal pneuma, or world soul, imagined by the ancient Greeks gradually became transformed into the idea of a universal ether, an invisible fluid of animating, pervasive energy that some believed contained the disembodied souls of the dead. We can see even among naturalists a predilection to sustain this notion of a life-giving energy. In the sixteenth century, William Gilbert, physician to Elizabeth I and an investigator with a special interest in lodestones, or naturally occurring magnets, coined the term "electrics"—derived from the Greek word for amber, a fossil resin that attracts small particles when it is rubbed—to refer to any substances with properties of attraction and repulsion. It seemed to Gilbert that inquiry into "electricity"—he coined that term, as well—might help to explain "animistic forces." Attraction and repulsion were themselves vital forces, he believed, making the earth, magnets, and other electrics in some way alive, even though they appeared inert. For the next three hundred years, "electrical philosophers," as some naturalists called themselves, would inquire into the nature of gravity, magnetism, and electricity, often using the terms interchangeably. What, they asked, was the constitution of this "subtle" or "imponderable"—meaning weightless—substance, and what was the connection of electricity to living matter?

Isaac Newton proposed that the "subtle and elastic" ether filling the universe also imbued nerves. While Descartes had pictured nerves as hollow tubes through which the vital spirit coursed, Newton suggested that nerves were solid filaments. "Is not Animal Motion perform'd by the Vibrations of this Medium," he wrote, "excited in the Brain by the power of the Will, and propagated from thence through the solid, pellucid and uniform Capillamenta of the Nerves into the Muscles, for contracting and dilating them?"[3] That question was central to the work of many eighteenth-century scientists, foremost among them Luigi Galvani and Alessandro Volta.

Galvani was a professor of anatomy and obstetrics at the University of Bologna, and a researcher on the role of nerves in muscular contractions, which in 1780, when his record of investigation on this matter begins, was a contested physiological problem. After experimenting with an apparently limitless number of frogs, in 1786, Galvani made an important new discovery. He hung frog legs across an iron railing, the spinal cords pierced by iron hooks. As the hooks touched the iron bar, "lo and behold," he wrote, "the frogs began to display spontaneous, irregular, and frequent movements. If the hook was pressed against the iron surface with a finger, the frog, if at rest, became excited—as often as the hook was pressed in the manner described."[4] Since the leg would kick even in the absence of external electrical stimulation, Galvani concluded that the frog's muscle served as a repository of animal electricity. More than a decade of research led Galvani to believe that artificial electricity and animal electricity were identical. Animal electricity, he maintained, was stored in the muscles, released by the will or, as he had shown in his experiments, by external stimuli, to cause movement.

Galvani's conclusions were publicized by stunning demonstrations given by his nephew Giovanni Aldini, also a professor at the University of Bologna and a popular showman. Frogs were suitable for laboratory experimentation, Aldini thought, but few spectators would come to watch frog legs twitch on a metal rod. To persuade his audiences of the reality of animal electricity, he would have to devise a more dramatic demonstration. This he did, by appearing onstage with a severed sheep's head to which he attached electrodes. As the current flowed, the dead animal's eyelids moved, the tongue shot out of its mouth, and it seemed to viewers that electricity had revived the animal's vital force. More shocking still, Aldini brought to the stage the bodies of recently executed murderers fresh from the scaffold, or their heads snatched from the guillotine. By exposing and electrifying the brain, Aldini managed to contort facial muscles into a grimace, to move the jaw, and to open and shut the eyes. The demonstration was grisly, yet it proved what many in his audiences wanted to be true: electricity animated living matter.

But desire was not the same as scientific proof, and Volta, a professor of experimental physics at the University of Pavia, was not convinced by Galvani's research. Like Galvani, Volta used frogs—along with his own ears, eyes, lips, arms, and tongue—to measure electricity. He maintained, however, that the electricity Galvani claimed to have discovered in the frog was instead a result of the contact of different metals and the moist organic matter of the frog's body. The metals caused a chemical imbalance that resulted in the production of the electricity that coursed through the muscle of the leg.

Yet Volta's certainty was undermined by the torpedo fish, which, like the electric eel, was a living example of animal electricity; in fact, the torpedo fish long had been used medically to treat patients suffering from epilepsy, for example, or heart failure. Volta was sure the electric eel and torpedo fish were an anomaly and not merely a more intense form of the electrical nature of other organisms. As he continued to experiment with the torpedo fish and to read others' work, he was struck by the description of the fish's electric organ: a column of membranes separated by a moist interlay. He created a model of that structure, but instead of using organic material, he used metals, which he knew would produce a stronger current. He alternated layers of silver and zinc, separating them by pieces of moist cardboard. Using brine as a conductor, he discovered that a pile of about forty of these pairs of disks emitted an electric shock similar to that of a torpedo fish. He called his new generator the "artificial electric organ." It became known as the voltaic pile, a device that could generate electricity continuously. Volta had no idea why the pile worked so consistently: "This perpetual motion may appear paradoxical and perhaps inexplicable," he admitted, "but it is nonetheless true and real, and can be touched, as it were, with the hands."[5] The voltaic pile, the first battery, had its roots in the investigation into animal electricity, yet it did not end the debate.

As persuasive as Volta was for some experimenters, his ideas did not capture the public imagination as deeply as Galvani's vitalism. Animal electricity seemed, simply, to explain the difference between life and death, and to account for consciousness, movement, and willpower. It offered a way to understand something as mysterious as sexual attraction and excitement, for which an electric shock seemed an apt metaphor. The idea that our bodies produced electricity shaped both scientific inquiry and popular beliefs in the nineteenth century, justifying electrotherapy and at the same time inspiring uncertainty and alarm about the creation and use of artificial electricity.

SPIRITUALISM

Vitalism was fundamental to spiritualism, the belief that the universal energy contained within the body—the soul—could persist when the physical being died. From the 1840s, when newspapers and magazines were filled with testimonies of ghostly rappings and phantasms, until the end of the century, Americans and Europeans from every class and educational background sat in dimly lit rooms where they witnessed tables turning, mediums communicating with the dead, apparitions materializing from behind gauzy curtains, and telepathic subjects guessing the contents of documents in sealed envelopes. They attended "mesmeric evenings" where personalities were "read" by men and women claiming special empathic powers. They were attuned, in their own daily lives, to tremors of inexplicable feelings that might portend disaster to someone far away whom they loved. Some became devotees of mesmerism, or magnetic sleep, a form of hypnotism caused, people believed, by the power of the mesmerist's animal magnetism. Despite assaults from science and medicine, mesmerism spread from eighteenth-century Vienna, where Franz Mesmer first treated patients, to nineteenth-century France and England, and then to America, where it was seen as a subversive path to illumination of both self and the vast invisible realm of the spirit.

Mesmer was born in 1734 in Vienna, where he studied and practiced medicine. Many patients sought his treatment for emotional symptoms: depression and irritability were common complaints from men; women suffered mostly from *vapeurs*, which caused fainting and nervous fits. Widely, if idiosyncratically, read, Mesmer attempted to combine Newtonian theories, as he understood them, with astrology and physiology. He believed that the sun, stars, and moon exerted an influence on tides, weather, and the special magnetic fluid that animated each individual's body. If this influence between ethereal forces and the body became blocked, illness would result.

As Mesmer considered the connection between universal energy and human health, he concluded that illness could be cured if the flow between the energized ether and the body were restored. One way to accomplish this restoration was through the use of magnets, which he believed harnessed ethereal energy and allowed it to enter the body. His early treatments involved a wide assortment of magnets applied wherever the patient felt pain. Soon, however, Mesmer's sense of his own power caused him to revise his therapy: if all humans contained animal magnetism, then magnets themselves might be irrelevant. After all, the physician, if he were powerful enough—as Mesmer was certain he was—should be able to vitalize a subject by allowing the strong force within him to emanate to the weaker ailing patient. While other physicians acknowledged to one another that their own sense of confidence or force of personality could contribute to their patients' healing, Mesmer publicly proclaimed that the physician's own body and mind could serve as therapeutic agents.

Mesmer's hubris, egotism, and theatrical form of treatment—including his preference for group therapy—scandalized the Viennese medical community; accused of fraud, he was forced to flee. In the late 1770s, he arrived in Paris, where he attracted a large following of the rich and curious. His apartment featured a sumptuous room redolent with perfume, which contained a *baquet*, or wooden tub, designed to serve as an electrical condenser. As patients stood around the circumference, grasping iron rods protruding from the tub, Mesmer would burst in dramatically, dressed in a lilac suit, flourishing a wand that served as the "pole" through which his own magnetic force flowed into his patients' bodies. For those patients, usually women, who suffered a "crisis" of emotions from the experience, Mesmer had private chambers where he could attend to each individually.

The transfusion of animal magnetism was heightened when, instead of a wand, the physician transmitted power with his own hands. In these treatments, the patient faced the physician, who held the patient's knees and feet between his own. Holding his hands about two inches from the patient's body, the physician passed his fingertips down the patient's arms, moving, as needed, to other parts of the body: from head to stomach, for example, or stomach to knees. The physician then placed himself in intimate, sexual proximity to the patient's body. A hundred years later, Freud, who was in the habit of administering massages and electrotherapy to his hysterical or otherwise neurotic patients, devised a cathartic therapy that involved pressing his hands against his patients' forehead while insisting that they tell him what burst spontaneously into their mind. Mesmer's treatment was no less cathartic for many patients, and no less dependent on the physician's belief in the strength of his own will.

Mesmer exploited popular assumptions about the connection of electricity and magnetism to sex, nowhere more blatant than in James Graham's Temple of Health, an extravagant setting for magnetic cures established in England in the 1780s. There, young women administered a variety of therapies to patients seated on magnetic chairs and thrones. The Temple's most notorious feature was a Celestial Bed, underlaid with more than a thousand magnets. The circulation of the magnetic emanations, Graham advertised, would ensure marital bliss. At fifty pounds a night, the experience was expensive, but apparently seductive for the wealthy clients who flocked to try it. Mesmerism proved equally seductive.

By the time Mesmer's followers began to practice in England early in the nineteenth century, mesmerism had become transformed from a group experience to a personal therapy enacted by one practitioner on one patient, and the infusion of energy resulted in the patient's falling into a trance, or "magnetic sleep." The idea that the vital spirit could be influenced by the powerful will of an empathic mesmerist attracted many people frustrated with conventional therapeutics for their physical or psychological pain.

Sufferers who turned to mesmerism often experienced the effects of mind on body. Samuel Taylor Coleridge, for example, hoped that mesmerism—rather than a narcotic drug—might offer him a way to control "some thought of Anguish" that could generate his physical pain.[6] But for Coleridge, mesmerism had another, equally compelling appeal: to serve as a technique for probing hidden, unexplored layers of mind, of personality, of selfhood. It was possible, Coleridge thought, that during the "morbid sleep" produced by mesmerism, "the Brain awakes, while the organs of sense remain in stupor."[7] This state of uninhibited consciousness might release the imagination in the same way that dreams did.

For a mesmeric trance to be effected, the strong-willed mesmerist needed to find a subject sufficiently sensitive and receptive. This receptivity did not imply passivity but rather a heightened sensitivity, a quality of which some people, especially poets, were exceptionally proud and which they described in electrical terms. Percy Shelley, as his cousin Tom Medwin described him, was excruciatingly "sensitive to external impressions"; he was "so magnetic" that he could become overpowered by others' sensual emanations. "When anything particularly interested him," Medwin reported, Shelley "felt a tremendous shivering of the nerves pass over him, an electric shock, a magnetism of the imagination."[8]

Percy and Mary Shelley long had been interested in the medical application of electricity. Even as a student, Percy Shelley experimented with electricity at home and in his rooms at Eton and Oxford. He administered electrical shocks to himself; during a thunderstorm, he attached a kite to a neighborhood cat; and he eagerly offered to electrify his sister Elizabeth in order to cure her of chilblains. Happily for Elizabeth, their parents intervened before Shelley could carry out his treatment. Both Percy and Mary had heard about, and perhaps even saw, Aldini's demonstrations of electrified, and apparently enlivened, corpses and animals; and they kept up with the medical community's response to animal magnetism, vitalism, and mesmerism, information that made its way into Mary Shelley's *Frankenstein.*

Other writers and poets shared their fascination. The language of electricity became a rich source of metaphor for sexual prowess and erotic feelings. But animal electricity and magnetism also suggested malevolence. Mesmerism implied emotional intimacy that could blur into feelings of being overpowered and violated by someone who could manipulate one's thoughts and invade one's private space. These disquieting notions about animal electricity formed a context for public understanding of artificial electricity. Although people flocked to demonstrations of electrical phenomena in lecture halls, although they permitted physicians to administer electrotherapy with the promise of cure, they were fearful about invasion by an invisible force.

SCIENCE

They were fearful, too, about the consequences of scientific investigation; fearful that the unquenchable desire to know would reduce the world to nothing more than empirical phenomena. The eighteenth century seemed a fury of electrical research in England (Joseph Priestley), America (Benjamin Franklin), and Europe (Pieter van Musschenbroek, inventor of the first electric condenser, or Leyden jar, and Abbé Jean-Antoine Nollet, chief electrician at the court of Louis XV). These men and scores of others—anyone, in fact, who could afford to buy a glass tube for producing static electricity—were able to set up a laboratory sophisticated enough to conduct experiments and to display their discoveries in stunning public demonstrations.

These displays were as likely to be conducted by amateur naturalists as by university professors, and they formed popular entertainment throughout the eighteenth and well into the nineteenth century. Audiences spanned social and economic classes, including royalty, academicians, merchants, lawyers, army officers, teachers, and shopkeepers, along with their wives and servants. In Paris, for example, private *musées* or *lycées* provided electrostatic generators so that demonstrators could produce such phenomena as lightning, shooting stars, a facsimile of the aurora borealis, an electrical model of the solar system, an electrified model insect so realistic it could fool the eye, a fiery "halo" shimmering around a subject's head as if evidence of beatification, or an electrified "Venus," whose kiss gave new intensity to the notion of sexual attraction.[9] Although the equipment needed to produce these visions was costly, it was possible for experimenters to reap considerable financial rewards. Electricity, noted Joseph Priestley, "furnished the means of subsistence to numbers of ingenious and industrious persons . . . who have had the address to turn to their own advantage that passion for the marvelous, which they saw to be so strong in all their fellow-creatures."[10]

Many groundbreaking researchers of the early 1800s—notably Hans Christian Ørsted, Humphry Davy, and Michael Faraday—found it possible to reconcile a quest for spiritual meaning with a search for physical, mechanical, or chemical laws. Ørsted's discovery that an electrical current can deflect a magnetized needle seemed proof of the unity of nature's forces. Davy, a leading light of the Royal Institution, considered chemistry to be "the most sublime and important of all sciences" for its potential to unveil God's will by discovering the "one law alone" that acted upon all matter: "an energy of mutation, impressed by the will of the Deity, a law which might be called the law of animation, tending to produce the greatest possible sum of perception, the greatest possible sum of happiness. The further we will investigate the phaenomena of nature," he proclaimed, "the more we discover simplicity and unity of design."[11] Faraday agreed: all forces were interdependent, convertible from one to another, and manifestations of the divine. Yet these scientists' conviction that in probing nature they were fulfilling God's will was undermined, as the nineteenth century progressed, by the technological applications of their discoveries, applications that spewed fumes, created clamor and glare, and seemed to have as their ends the financial enrichment of businessmen rather than the enlightenment of humankind.

As science pushed aggressively to uncover Nature's secrets, the public encountered, in magazine essays, newspaper articles, and fiction, disturbing characterizations of scientists as heartless, godless transgressors. If scientists accumulated knowledge, they did not necessarily impart wisdom; as Coleridge so trenchantly observed, for "real utility" and true enlightenment one would be better off consulting Shakespeare or Milton.[12] It is no surprise that Edison crafted a public image as an inspired inventor rather than a scientist; at the height of his popularity, the press portrayed him as a wizard: the wielder, and therefore the protector, of occult powers. But as Edison became the subject of incessant publicity, that image changed: increasingly, he was associated with scientists, whose motives were suspect, or with businessmen, whose interests were purely self-serving.

Throughout the nineteenth century, electricity was temptress and seducer, feared and coveted, a force that could animate life or inflict death. Electrification threatened a public who treasured shadows and secrets, and who yearned for a universe in which "wild facts" preserved a feeling of wonder. What electricity generated most pervasively was anxiety.

We describe ourselves in electrical terms: wired, shocked, electrified, charged, in need of recharging. We know some "human dynamos" and "powerhouses." We plug into new ideas. We hold beliefs about the electrical activity of mind and body, about human emanations and "auras," that bear upon the medical treatments we seek and in which we place our faith. Electricity is still part of the medical arsenal (in treating mental illness and heart disease, and in physical therapy), and it also surfaces in self-help therapies, such as magnetic insoles, bracelets, and belts. But we have inherited another legacy as well: a pattern of responses to new technologies that allure, threaten, and urge us to reinvent our sense of self. *Dark Light* is about that legacy: the intellectual and emotional revolution that occurred as ordinary men and women, eagerly and anxiously, grappled with the greatest new idea of their time.

PART I

Wonders

CHAPTER 1

Working Great Mischief

Morse/Vail telegraph key, 1844.
The Smithsonian Institution

The electric telegraph is the miracle of modern times . . . a man may generate a spark at London which, with one fiery leap, will return back under his hand and disappear, but in that moment of time it will have encompassed the planet on which we are whirling through space into eternity. That spark will be a human thought!

The Times, London, October 1856

When he was twenty-three, Samuel Finley Breese Morse wrote to his parents from England, where he had been studying and practicing art for two years, defying their wishes that he become a bookseller at home in Massachusetts. Morse had just received an Adelphi Gold Medal for a statuette of Hercules, and *Dying Hercules*, his painting of the same subject, recently had been exhibited at the Royal Academy. Despite these achievements he was well aware, as were his parents, of the challenges that any artist faced simply in making a living. If those challenges were not quite Herculean, surely they seemed, at the time, heroic. "I need not tell you what a difficult profession I have undertaken," he wrote. "It has difficulties in itself which are sufficient to deter any man who has not firmness enough to go through with it at all hazards, without meeting with any obstacles aside from it."[1]

In 1815, Morse returned to America, where he believed he would face even more obstacles as a professional artist than in Europe. "I should like to be the greatest painter *purely out of revenge*," he proclaimed.[2] Awaiting recognition of his greatness, he opened an art studio in Boston; but clients did not flock to his door, and he resorted to a path he had hoped not to take: cultivating a career as a portrait painter. By the time he married in 1818, he had earned considerable renown, attracting commissions from politicians, college presidents, statesmen, and other public figures. Along the way, he also dabbled in inventing: he devised a fire engine water pump, which unfortunately failed; and a marble-cutting machine, which he could not patent because it infringed upon another inventor's design.

In February 1825, Morse was in Washington, painting a portrait of General Lafayette, when his wife, at the age of twenty-five, suddenly died a month after giving birth to their son. By the time Morse received the shocking news and returned to New Haven, where she had been living with his parents, the family already had buried her. This was not the first time that delay in receiving a message had caused Morse distress: while he was in England studying, letters to his family went astray, or their letters to him failed to arrive, generating suspicion and worry for all parties.

After his wife's death, followed by his father's death the next year and his mother's in 1828, Morse fell into a depression. Like many of his contemporaries, he sought solace in travel abroad, and in 1829, leaving his children with relatives, he sailed to Europe to paint, study, and revive his spirits. Three years later, he sailed home. The return trip, as it turned out, changed his life. During the six-week journey, he had several conversations with other passengers about electromagnetism, a phenomenon reported about frequently in newspapers and magazines. Although electricity and magnetism had been thought of as separate forces, recent discoveries by Hans Ørsted, Michael Faraday, and the American physicist Joseph Henry proved that the forces were interrelated; electrified wire acted exactly as a magnet—able, for example, to deflect the needle of a compass—and electrification could enhance a magnet's strength. This discovery was exciting news for those who yearned to prove the unity of natural forces; it was also exciting for inventors tinkering with the possibility of sending signals over long distances by interruptions in electrical current.

Morse had only a limited background in electricity. He had attended some lectures on the subject when he was an undergraduate at Yale, and in 1827, he had heard Columbia professor James Freeman Dana lecture on electricity and electromagnetism at the New York Athenaeum. Interested in what Dana had to say, Morse befriended him for further conversations. These talks with Dana and later with his fellow passengers, besides introducing Morse to recent investigations into electromagnetism, most likely made him aware of prototypes of the telegraph developed since the middle of the eighteenth century. Volta's invention of a chemical battery in 1799 and new reports of the effect of electricity on magnets generated excited speculation that a breakthrough in telegraphy was imminent.

Inspired by his shipboard conversations, Morse returned to his cabin with his imagination fired. Although he had never before dabbled in electricity, he felt convinced that electromagnetism could be harnessed to facilitate communication. Soft wire bent into a horseshoe shape, Morse believed, could be magnetized by sending a galvanic current through a coil wound around the iron; the coil would lose its magnetism when the current was suspended; these alternations could produce marks on a paper, if a stylus and paper roll somehow were attached to the mechanism. Fascinated by this idea, Morse drew a design for a Recording Electric Magnetic Telegraph, along with a code in which numbers corresponded to dots and dashes. These numbers, he conjectured, then would correspond to letters of the alphabet, and the code would result in a special telegraph dictionary.

Morse arrived home on November 15, 1832, enthralled by the idea of the telegraph, yet facing the reality of having to make a living. A widower with young children, Morse knew that he would have to continue portraiture in order to provide for his family. In 1835, however, a fortunate occurrence solved both of his problems: he was appointed Professor of the Literature of the Arts and Design at the University of the City of New York. This position gave him enough income so that he could retreat from portrait painting and devote himself fully to his invention. Within a year, working dutifully in an old building in Washington Square, where the university had given him studio space—and where Morse also ate, slept, and taught art—he had constructed a model of his telegraph.

At first glance, it looked like a motley assemblage of wood and metal. Morse fastened a canvas frame to a tabletop. He found wheels from an old wooden clock to carry a strip of paper forward, over and under three wooden drums. He suspended a wooden pendulum from the top piece of the frame, with a pencil at the end which came into contact with the paper. To provide power, he attached an electromagnet to a shelf across the frame, connected by a short circuit of wire to the terminals of a battery. Odd as it appeared, it worked.

Early in 1836, Morse demonstrated his instrument to a few trusted men, among them Leonard Gale, a professor of science, and Alfred Vail, a young entrepreneur with an interest in new technology; both were enthusiastic enough to back Morse financially. With Gale's advice, Morse devised a sequence of circuits, or relays, which increased the distance an electrical impulse could travel and thereby enable messages to be sent to a receiver miles away. Morse was elated about the potential of his invention; he was sure, he said, that "magnetism would do in the advancement of human welfare what the Spirit of God would do in the moral renovation of man's nature; that it would educate and enlarge the force of the world."[3] This extravagant claim would be echoed by future nineteenth-century inventors of other technologies: if humans dared to manipulate nature, they must do so for the good of humankind. Education was one proper effect of telegraphy, moral uplift even better. The prospect of financial gain clearly was not to be admitted publicly.

Yet Morse realized, indeed was counting on, the possibility of such gain, and he was worried that telegraphy was likely to be coveted by others. Morse knew, of course, that many inventors were at work on the same project, both in America and abroad: "before it can be perfected," he wrote to a friend, "I have reason to fear that other nations will take the hint and rob me both of the credit and the profit."[4] And other nations were not his only threat: responding to rumors of Morse's invention, one shipboard companion with whom Morse had discussed electromagnetism claimed to be the true inventor. In the fall of 1837, fearing a repetition of his failure to gain a patent on his marble-cutting machine, Morse immediately sought and received patent protection for his telegraph.

A few months later, Morse began to move quickly both to firm his reputation as inventor and to raise funds for further refinement of the device. He demonstrated the telegraph at Philadelphia's prestigious Franklin Institute and before the Committee on Commerce of the House of Representatives. It seemed to Morse both logical and prudent that the government own rights to telegraphy. "It is obvious," he wrote to Francis O. J. Smith, a congressman from Maine who clearly was impressed with the invention, "that this mode of instantaneous communication must inevitably become an instrument of immense power, to be wielded for good or for evil, as it shall be properly or improperly directed... even in the hands of Government alone it might become the means of working vast mischief to the Republic."[5] He offered the government exclusive rights to telegraphy; the government declined.

Not fully defeated, Morse asked for $30,000 to fund further research and development, including construction of a telegraph line between Baltimore and Washington. This bill, sponsored by Smith, who, unethically for a public servant, had financial interests in Morse's invention, won narrow approval. But congressional funds were not disbursed; the Panic of 1837 had left the government, and the economy as a whole, financially depressed. Once again, Morse found himself without the resources to continue his research and development. This time, even his investors Gale, Vail, and Smith did not offer additional support.

Morse decided that a successful public demonstration of telegraphy would serve him well, and in October 1842, he rented a rowboat and rower, setting off from the Battery toward Governor's Island. The boat carried a cable insulated with pitch, tar, and rubber, and this Morse gradually dropped into the waters off Manhattan. But the next day, when he attempted to send a message between Castle Garden and Governor's Island, his signal inexplicably failed. During the night, he later learned, a ship discovered the cable and, not knowing what it was or why it was there, cut it.

Once again, he petitioned Congress for funds. This time the bill generated contentious debate that revealed both the public's lack of interest in telegraphy and their suspicion about electricity itself. One of Morse's most vociferous detractors proposed an addendum requiring that half of the funds be devoted to exploring mesmerism, "to determine how far the magnetism of mesmerism was analogous to that to be employed in telegraphs."[6] Evoking the connection between mesmerism and telegraphy was not entirely facetious: sending thoughts through wires seemed, at the time, just as foolish and arcane as sending thoughts, without wires, from one human being to another. Both projects had the aura of the occult; both projects, as one magazine article put it, seemed "impracticable and Utopian."[7] Morse's bill, without the amended ridicule, passed by only six votes, the money was disbursed immediately, and Morse began to construct the first telegraph line, between Baltimore and Washington. The process—frustrating, exhausting, fraught with delays and mistakes—took more than a year.

On 24 May 1844, Morse sat in the Supreme Court chamber in Washington and tapped out the message he believed most fitting for the occasion: "What hath God wrought!" This first public telegraph message echoed the sentiments held dearly by others: humans were the instruments of the Supreme Inventor; everything revealed to humankind was done so as part of a grand design. Why else, Morse reflected, did the same idea occur to so many inventors at the same time? Surely because "the Great Author of all good, the Giver of every great gift to the world, intends when such a boon is bestowed that He first and prominently shall be recognized as the Author... Man is but an instrument of good, if he will fulfil his mission."[8] Articles repeatedly referring to the telegraph as a miracle underscored this connection to the divine. Morse was but a "co-laborer" in God's vast workshop.

In the Baltimore railroad depot, Alfred Vail received Morse's humble message instantly. Witnesses at both ends were impressed; and yet, when Morse again offered the government rights to telegraphy, the government again declined. Postmaster General Cave Johnson, who had his own interests in minimizing the telegraph's impact on communication, refused to approve funding "on the ground that, however beneficial it might be as a private enterprise, and however advantageous to the Government in the rapid transmission of intelligence, yet it could never become a paying concern." Johnson accurately read the mood of the nation. Even when the Baltimore–Washington line opened for public use, the public came only to observe. At a charge of one cent for every four characters transmitted, the telegraph, in the first months it was available to the public, earned just pennies a day.

A MAGNETISM OF THE IMAGINATION

The public's lackluster response to the new technology resulted in part because they failed to understand how it worked. People who came in with messages written on a piece of paper expected that the paper itself somehow could be transmitted through a cable. Others thought that objects—such as a gift—could be transmitted. Many who did understand that electricity had been manipulated to transmit writing over distances conflated that idea with the transmission of thought over distances. Like the congressman who suggested that government funds be appropriated for an investigation into mesmerism, some people suspected that telegraphy had a direct connection to the workings of the mind.

Lack of privacy disturbed some potential users: they did not like the idea that their message would be witnessed and intercepted by strangers. In Henry James's story "In the Cage," published more than fifty years after Morse first introduced his invention, a telegraph operator proves clever enough to draw inferences about her customers' lives from their elliptical messages, and she "quite thrilled herself with thinking what, with such a lot of material a bad girl would do." Her knowledge, she decides, could be an effective "purchasing medium."[9] If blackmail was not likely for most message senders, certainly invasion of privacy was inevitable. In England, inventor William Cooke suggested that the deaf and dumb might make excellent telegraph operators since they were "accustomed from their infancy to abbreviate as far as possible, by signs, in their symbolic language"[10] and since, presumably, they could not convey those messages to the hearing and speaking public.

Cost also proved a significant factor in resistance to telegraph use, just as it would for future technological innovation. Businessmen whose profits depended on the quick transmission of prices for stocks and goods seemed the most likely beneficiaries. But for ordinary users, the benefits did not justify the price. James's protagonist "often gasped at the sums people were willing to pay for the stuff they transmitted—the 'much love's, the 'awful' regrets, the compliments and wonderments and vain, vague gestures that cost the price of a new pair of boots."[11] In contrast to the wealthy telegram senders who paid such sums in James's story, few thought that their messages were important enough that they needed to be sent and received with urgent haste.

In the months after Morse sent his first public message, the most enthusiastic publicizer of the telegraph was the press, which saw the potential of instant messages for the enhancement of reporting and the selling of daily newspapers. The new year of 1845 opened with the *Albany Argus* proclaiming that the telegraph offered "a stride in the march of intelligence of no ordinary importance. It is one of those triumphs of the arts of peace that knit our people in closer relations of union and brotherhood. The Magnetic Telegraph annihilates distance." Whatever the government decided in Washington, whatever words the President uttered, whatever European news arrived by ships docking in Boston or New York could be known instantly in any and all cities connected by telegraph.[12] On July 18, 1846, the *Baltimore Patriot* announced "LATEST NEWS!!!" from Europe transmitted to the paper "By Magnetic Telegraph" just hours after the steamship *Cambria* docked in Boston; and the paper's dispatches from Washington also carried the headline "By Telegraph." But this excitement by the media proved not to be infectious in other quarters, and Morse still could not raise money.

In the spring of 1845, a despondent Morse began conversations with Amos Kendall, a lawyer and former postmaster general, who offered to take on the role of business agent in exchange for three-fourths interest in Morse's patent. The remainder was held by Francis Smith. Yet even Kendall failed to persuade the government to finance telegraph lines nor could he raise much private capital. In the end, he put up some money himself for a New York to Washington line; Smith financed a connection between New York and Boston, and a relative of his agreed to pay for a line from Washington to New Orleans. Two men from upstate New York agreed to finance a line from Buffalo to Manhattan and from the eastern coast to the Great Lakes. With this support promised, Kendall devoted his efforts to funding his own company to build a line from New York to Philadelphia, the first stage in the New York to Washington connection that he planned. Raising the needed $15,000 was arduous, but by May 1845, he had found twenty-five men willing to invest a few hundred dollars—only six invested a thousand—and the Magnetic Telegraph Company was, at last, incorporated.

ACCELERATIONS

In England, the telegraph became an important technology for railroads, replacing the semaphore as a signaling device that could more reliably prevent accidents, especially in fog. In America, however, the telegraph was taken up mainly by businesses, which saw how rapid communication about prices and orders had the potential of increasing profits; the technology became a business in itself. By 1846, Morse's year-old Magnetic Telegraph Company had eight rivals with lines from Portland, Maine, to Savannah, Georgia, and westward as far as Chicago. Rival companies meant that there was no unifying plan for expansion, no common standard for the type of wire used for telegraph lines, and unbridled competition for customers. Unreliability was the technology's major problem: poles were flimsy; thin copper wires easily broke; insulators, shaped like glass knobs, served as target practice for sportsmen; if a company tried to insulate its wires with tar or wax, they soon attracted a coating of insects.[13] Operators were not always available, and some users accused the companies of bias about whose messages were sent first.[14]

Just as technical shortcomings caused stress among business users, so did the pressure to make quick decisions to maximize profits and to act instantly in response to a continuous onslaught of information. Outside of the marketplace, men and women had little experience with the reality of the telegraph; instead, they were subjected to contradictory views in newspaper and magazine articles, some complaining about the technology's problems, many more proclaiming its potential to transform social, political, and cultural life. These articles offered claims about telegraphy even more hyperbolic than Morse's own. The telegraph would enhance conversations among world leaders and enable them to have instant access to one another. In fact, the telegraph might generate "a common language of the world" because of the ease and frequency of communication. Better communication, the press insisted, necessarily would lead to better understanding: wars would be a blight of the past. The telegraph would be a force for democratization: "It is an institution for the people," one minister proclaimed at the First Church in Boston after the transatlantic cable was laid in 1858. "Men who talk together daily cannot hate or disown one another."[15] But this idea that the telegraph could make individuals feel like citizens of the world suggested a new burden of responsibility to a community larger than one's hometown; an enlarged sphere of citizenship required an understanding of international needs and problems far deeper than what could be gleaned in a "vivid flash" of a message.

Escalation in the content and kind of information available to the public implied other changes in daily life. Journalists suggested that the postal system would shrink by half, with mail becoming "a still tolerably convenient but antiquated and dyspeptic institution"; railroads, too, would shrink, as businessmen conducted transactions by telegraph rather than in person.[16] Even language would change. "I think the habit of writing by telegraph will have a happy effect on all writing by teaching condensation," Emerson remarked.[17] But condensation might be less than happy if it meant unimaginative conformity. With the telegraph, words had a monetary value: the greatest amount of information must be conveyed in the fewest words. "Language is but the medium of thought," one writer commented, and thought must be pared down to its essentials.

"Every useless ornament, every added grace which is not the very extreme of simplicity, is but a troublesome encumbrance." Language would become homogenized, standardized, and unreflective of the writer's or speaker's personality. "The Telegraphic style, as we shall denominate it, for the benefit of all future writers upon rhetoric," this journalist proposed, "is also terse, condensed, expressive, sparing of expletives and utterly ignorant of synonyms. From its subject matter it has little to do with beauty or grace." Soon, newspapers and even personal correspondence would come under the influence of this "perspicuity."[18]

Most persistent, appearing in article after article, was the exuberant proclamation that the telegraph had the power to "annihilate space and time"[19]—a power that seemed at once enticing and threatening. The advantage of killing time seemed suspicious. When information came too quickly—as business users already knew—one would not have time to digest, consider, and contemplate meaning. The telegraph's implication of urgency meant that people would be forced to contrive a quick response, and haste, as the saying goes, was likely to mean waste. Although newspapers believed it would be a great advantage to the country if presidential election results could be conveyed instantly, articles intended to whet the public's appetite implied, instead, that the telegraph would bring nothing but bad news. A "wife in Louisville" could be informed that her husband was run over by an omnibus in New York and could be kept updated, hour by hour, about his injuries; the public would know instantly if "a treacherous villain seduces his friend's wife and escapes, with as much of her husband's property as they can easily carry"; bank robberies and even forgeries would become publicized far from the site where they occurred to people for whom these events were of no pressing concern.

An onslaught of information about crimes and mischief, conveyed by telegraph, could be disseminated without "a single second of time intervening between the event and the universal diffusion of the intelligence."[20]

For Nathaniel Hawthorne, alert to the potential impact of technology on daily life, apprehending criminals was not the proper use of the "almost spiritual medium" of the telegraph; instead, as his protagonist Clifford proclaims in *The House of the Seven Gables*, it "should be consecrated to high, deep, joyful, and holy missions." Lovers would send messages chronicling their "heart-throbs": "I love you more than I can!" "I have lived an hour longer, and love you twice as much!" A death, a birth: these announcements are worthy of electrical messages, not, says Clifford, "the enlistment of an immaterial and miraculous power in the universal world-hunt" for robbers and murderers.[21] But whether the "electric thrill" was used for the prosaic or sublime, more news would surge into every community.

The eradication of space implied other problems: individuals would be drawn closer to the rest of the world, to be sure, but also would be pushed into proximity, perhaps unwanted and uncomfortable, with strangers. If, as one journalist proposed, American cities and towns all became suburbs of New York, the center of telegraph communication in the east, then place would lose its regional identity and integrity: one would not be sure if neighbors shared common perceptions and beliefs. "One of the evils of this age of railroads and telegraphs," a magazine writer commented years later, "is that we are forced to know people as they are."[22] The telegraph presented a chance to enlarge one's world, to move beyond the familiar and often homogeneous environment of town or city, but the chance for expanded experience felt dangerous. If, as the *Albany Argus* excitedly predicted, the telegraph would "literally render our people one family," that impending kinship implied a darker side: intrusion and invasion. Individuals would need to set boundaries between themselves and those at the other end of an electrical wire. Telegraph users worried about privacy and exposure, and yet as they did, the potential for connection made them aware of a yearning to transcend personal boundaries. If telegraph cables were the nerves connecting various parts of the social and political body, if telegraph stations were the brain to which impulses, information, and "intelligence" were transmitted, it could offer unimagined possibilities for sympathy and empathy.

INSULATION

The telegraph raised questions about risk that seemed unrelated to the technology itself: What were the risks of increased knowledge, not just of news and information but of other people? What were the risks of connecting with another person through electrical impulses conveyed by wires? What kind of relationship was possible through interactions by brief messages? What was one's responsibility to people who were voiceless and faceless? In a world teeming with news, what were the risks of insularity?

These questions are addressed in "An Evening with the Telegraph-Wires," a story that appeared in *The Atlantic Monthly* in 1858, when, because of the laying of the Atlantic cable, telegraph technology once again emerged as front-page news. The story's narrator agrees to serve as a subject for his cousin Moses, who possesses "the mesmeric gift" and is eager to demonstrate his talents on women with nervous headaches or, in fact, on anyone who will sit still long enough to become entranced. Moses is not quite successful with his cousin, however, and after half an hour of effort, both agree to end the experiment. Although Moses makes the requisite "reverse passes," the narrator leaves the session feeling "strangely excited," and he decides to take a walk to work off this nervous energy. During his walk, he suddenly is inspired to climb a tree; sitting among the branches, he hears "a low vibrating note" and sees several telegraph wires strung beside him. The vibrations convince him that a message is being sent at that very moment, and he decides, as a kind of scientific experiment, to see if he can "hear or feel the purport of a telegraphic message, simply by touching the wire along which it runs!" The experiment is a success: after a few short jerks of energy, the narrator realizes that his brain has recognized a telegraphic message.

Although the message conveys nothing more than stock prices, the narrator's achievement sends him into a reverie: "What a thing this discovery of mine would be for political conspirators," he thinks; "instead of the tyrant hearing the secrets of the people, the people hearing the secrets of the tyrant!" Another incoming message reveals the "semi-articulate sob" of a mother imploring her husband to rush home before their young daughter, mortally ill, dies. The narrator then finds that he does not merely intercept "intelligence" but can connect emotionally with the mother's sorrow. Something about the medium, he decides, allows this intimacy. "Strange!" he reflects. "One reads such an announcement in a newspaper very coolly;—why is it that I can't take it coolly in a telegraph dispatch? We can read a thing with indifference which we hear spoken with a shudder,—such prisoners are we to our senses!"

After listening to a few more messages, the narrator feels that he has "expended enough sympathy, for one night" and goes on his way. He is convinced that he has become a medium, a talent that does not surprise him. "After all," he considers, "why should it be thought so improbable, in this age of strange phenomena, that the ideas transmitted through the electro-magnetic wire may be communicated to the brain . . . Is it not reasonable to suppose that all magnetisms are one in essence?" Animal magnetism, the basis for mesmerism, as he understands it, surely is the same as electromagnetism, the basis of the telegraph. "I put these questions to scientific men," he says; "and I do not see why they should be answered by silence or ridicule, merely because the whole subject is veiled in mystery."

The narrator's greatest discovery, however, is not the unity of magnetisms but the need for human connection. As Emerson once admitted in his journals, he felt an insatiable need "to know if yellow be yellow, & grass grass to another eye."[23] The telegraph, promoted as a means of annihilating the time and space between individuals, offered the potential of connecting each individual to the joys and sorrows of another. True, the narrator concedes, too much connection "would interfere with our freedom and our happiness." True, he advises, a certain amount of "insulation is necessary, that individual life and mental equanimity may go on." But he implies that there is already too much insulation, too little connectedness.[24] Maintaining one's insularity implies selfishness, stubbornness, and fear. Yet insularity can offer protection from the risks of suffering emotional or intellectual shocks. The telegraph, it seemed, presented humanity with a difficult choice.

COMMON PROPERTY

The premise of "An Evening with the Telegraph-Wires" drew upon the popular conflation of artificial and animal electricity, a notion perpetuated in newspaper and magazine articles. These writings confirmed the pervasive belief that the electricity flowing through telegraph lines was the same force that moved human muscles. An article in the staid *North American Review*, published thirteen years after the Magnetic Telegraph Company was established, offers the typical and often repeated view that the telegraph is energized by a force familiar to every reader:

> It has been the medium of all communication between mind and matter, brain and muscle, brain and brain; and in the phenomena of mesmerism and of pseudo-spiritualism, there is at least some reason to believe that, along air-lines and for indefinite distances, thoughts and words are sent with as unerring fidelity as marks their transmission on the artificial lightning-path. By the connection now established between distant cities and opposite hemispheres, we have but arrested, for a special subdivision of one among its many departments of service, a force which throbs from zone to zone, leaps from sky to earth, darts from earth to ocean, courses in the sap of the growing tree, runs along the nervous tissue of the living man, and can be commanded for the speaking wires simply because it is and works everywhere.[25]

Since electricity was everywhere in nature, waiting to be arrested and exploited by humans for their own purposes, it was easy to believe that the telegraph marked the beginning of a larger and more pervasive electrical revolution. Electricity was "common property," a "cosmical element of the universe" available to everyone, and its power was likely to be extended. Electrolysis of water, it was believed, would yield a clean-burning gas suitable for cooking and lighting. "Water contains electricity in enormous quantities," one writer exclaimed in an article titled, ecstatically, "Touching the Lightning Genius of the Age." "Faraday says a single drop of it holds the lightning of a thunderstorm!" When this flammable gas was harnessed, there would no longer be a need for coal mining and deforestation. Such predictions might seem incredible, the writer admitted; but, he advised readers, "This is not an age to be astonished at any thing; and Science now-a-days asks as large a faith as Superstition did formerly."[26] Faith, rather than understanding, characterized popular notions about electricity, and if one believed that a drop of water contained lightning, it seemed no less likely that human tissue did, too. "Call it by whatever name," Emerson noted in his journal, "we all believe in personal magnetism, of which mesmerism is a lowest...example."[27] And the telegraph, as Morse's early antagonist noted, was merely another manifestation of the magnetism that coursed through all of nature and each living body. As the prestigious journal *Nature* put it, the telegraph was capable of "transmitting power of the electric fluid over the metallic nerves of speech."[28]

The telegraph served as more than an instrument for sending brief messages quickly: it was a portent of future manipulation of nature, manipulation that, some people thought, might incur Nature's wrath; it was an intimation of vast social, cultural, and political changes; it threatened to be a source of invasion and intrusion into private life; it threatened to change one's sense of time and space; it threatened to overwhelm its users with information and insist on their rapid response; it urged people, even those who had never used a telegraph, to examine their beliefs about the relationship between artificial and animal electricity; it aroused superstitions. In short, it evoked responses that Morse could not have predicted when he retreated to his ship's cabin during his journey home.

Morse's world, of faith and superstition informing technological change, of science and religion both offering compelling evidence of the miraculous and transcendent, was the world into which George Beard was born in 1839 and Thomas Edison in 1847. Electricity featured in their lives, even as boys: Edison used batteries to power the makeshift instruments that he constructed in his parents' basement; and the adolescent Beard administered electrical treatments to himself for his recurring digestive problems and depression, symptoms caused largely by his struggle to reconcile his family's religious beliefs with his own. Decades later, electricity would prove central to both men's careers. But when they were young, it was, as for most of their contemporaries, a tantalizing and enigmatic force.

"When I was a little boy, persistently trying to find out how the telegraph worked and why," Edison recalled, "the best explanation I ever got was from an old Scotch line repairer who said that if you had a dog like a dachshund long enough to reach from Edinburgh to London, if you pulled his tail in Edinburgh he would bark in London. I could understand that. But it was hard to get at what it was that went through the dog or over the wire."[29]

CHAPTER 2

Beneficence

Thomas Edison, about age 27.

U.S. Department of the Interior, National Park Service, Edison National Historic Site

Who shall ever say *Impossible* again? Henceforth, if a thing is desireable, it is really practicable, & whatever you have dreamed of, go instantly & do.

Ralph Waldo Emerson,
Journals of Ralph Waldo Emerson, vol. xvi, 1868

Is Electricity Life?" asked Henry Lake in the *Popular Science Monthly* in 1873. The question was hardly new, and neither was his answer—an enthusiastic and lyrical "yes." Like generations of writers who claimed that electricity was the essential animating force, Lake characterized electricity as nurturing, benign, generous, and protective. "It is the very soul of the universe," Lake declared. "It permeates all space, surrounds the earth, and is found in every part of it... It is naturally the most peaceful element in creation. It is eminently social, and nestles around the form it inhabits. Unlike many human specimens, it never desires to keep all its good to itself, but is ever ready to diffuse its beneficence."[1] No one benefited more from the generosity of electricity than Thomas Edison.

Thomas Alva Edison was born in Milan, Ohio, on February 11, 1847, the seventh child of Samuel and Nancy Edison.

Among many families who left Milan looking for better economic opportunities, the Edisons moved to Port Huron, Michigan, where, at the age of eight, Tom attended school for the first time. Willful and uninterested in rote memorization, he proved himself a difficult student, and his mother decided to give him lessons at home. Like many boys his age, Tom set up a laboratory in the basement of his house, where he proceeded to learn chemistry and physics on his own with whatever supplies and chemicals he could gather from scrap heaps or buy with pocket money. Sometimes he tested his chemical concoctions on other children, with unfortunate, though not deadly, results. His mother made him label all his chemicals "poison" to avoid any further mishaps. He also devised his own telegraph, stringing a line through the woods around his house.

Tom's laboratory kept him so fully occupied that he had little time for friends, sports, or play. He was willing to leave his experiments to earn money for supplies, and at the age of twelve took a job as a newsboy on the Grand Trunk Railroad that ran from Port Huron to Detroit. Each day Tom left his home at dawn, returned at eleven, and spent layovers wandering alone in Detroit. Yet he managed to pursue his experiments in a makeshift laboratory that he set up in the train's baggage car.

Early in this period, Tom Edison became considerably deaf. His biographers have speculated about the cause: a previous puncture of an eardrum, a childhood bout of scarlatina, an accident or blow to the ears, or a congenital problem that worsened. By the time he was thirteen, he could hear only when shouted at, unable to participate in ordinary conversation, unable, he recalled, to hear birds sing. Deafness increased his sense of isolation; he began to spend his layovers at the Detroit Free Library, where he read indiscriminately and hungrily: science, literature, encyclopedias, and, by his own testimony, philosophy.

In April 1862, news of the devastating Battle of Shiloh reached Detroit by telegraph, attracting huge crowds as soon as the casualty figures—some 60,000 killed and wounded—were posted on a chalkboard at the station. Edison had the idea that if such news were telegraphed ahead to smaller stations, the sale of newspapers would rise dramatically. He struck a deal with a Detroit telegrapher, who, in exchange for a few months' subscription to some popular magazines, agreed to telegraph to all the stations whatever news he received. Fifteen-year-old Tom then proceeded intrepidly to ask Wilbur Storey, managing editor of the *Detroit Free Press*, for a thousand papers on credit. Perhaps because of Tom's guileless audacity, Storey agreed. Tom and a friend lugged the papers to the train, folded them, and loaded them. At the first station, because the headlines already had been publicized by telegraph, Tom sold thirty-five papers instead of his usual two: "then I realized," he said later, "that the telegraph was a great invention."[2] At the next stations, with even greater demand for papers, Tom decided he could raise the price from a nickel to ten cents, and then to twenty-five. He sold out his entire supply. "If it hadn't been for Wilbur F. Storey," he said later, "I should never have fully appreciated the wonders of electrical science."[3]

Edison's enterprise was motivated less by his interest in telegraphy than in money. Ever since he began working on the train, he had supplemented his earnings by buying produce from farmers along the route and selling it in stands or by hiring other boys to sell magazines, tobacco, and candy along the way. Telegraphy was a means to an end: quick profits. Inspired by his success in selling news, he set himself to learning telegraphy and printing—although correct spelling eluded him—purchasing a small hand press and churning out the *Weekly Herald* from the baggage car that once had served as his laboratory. This project lasted only a few months, since in the summer of 1862 a chance occurrence resulted in a significant opportunity.

One day when he was sitting in the telegraph office at Mt. Clemens station, Tom Edison noticed that the station agent's three-year-old son had wandered onto the train track. He rushed out to grab the child moments before a freight train came along. The boy's mother fainted, Edison recalled, and his father showed his gratitude by offering to teach Tom telegraphy. Tom moved in with the family, where he contributed to the cost of his food, and within a few months, working diligently for eighteen hours a day, he became so proficient that he was able to set himself up in a small office back in Port Huron. Edison's skill, coupled with his ability to work long hours with little sleep, put him in demand. During the war years, when telegraphers were in short supply, Edison found work easily. He moved from job to job, sometimes dismissed—apparently for insubordination or for tinkering with the equipment—sometimes looking for higher pay, often just because of his restlessness. By the fall of 1864, he was employed in a Western Union office in Cincinnati.

Living in boarding houses that were so derelict they would have been "a paradise for an entomologist," Edison was one among many brash and aspiring young men suffering from a "long estrangement from money."[4] He fiercely resolved to end that estrangement. At the age of twenty, he discovered Michael Faraday's *Experimental Researches in Electricity* and was inspired by the scientist's perseverance in experimentation and by the sheer volume of Faraday's discoveries. Faraday, who rose to eminence despite little formal education and who assiduously nurtured his public image as a scientific authority, became an exemplar. Edison, after all, knew that he was no theoretician, and college-trained scientists intimidated him. "Was Voltaire the inventor of the Voltaic pile," he once wrote ingenuously to the editor of the *Telegrapher*, "or was it Volta?" In a Boston telegraph office, where other operators were studying to enter Harvard, Edison was determined to prove that he was just as smart. "They paraded their knowledge rather freely," he remembered, "and it was my delight to go up to the second-hand book stores on Cornhill and study up questions which I would spring on them when I got an occasion."[5] What he lacked in theoretical knowledge, however, he made up in keen business sense, sheer aggression, and imaginative genius.

Edison began as a tinkerer with an eye for simplicity. Faced with an infestation of cockroaches at the Western Union office, he invented a device for electrocuting the insects that won him a humorous notice in a Boston newspaper. He could fix a machine to make it run more efficiently; he could design a machine to fill what he saw as a need. The first patent he ever filed, in October 1868, was for a vote-recording machine that could quickly and accurately tally votes in Congress. The reception of his idea recalls what Morse encountered twenty-five years earlier: the device, a congressional committee chairman told Edison frankly, would get in the way of filibustering.

The vote recorder was one of many projects that Edison worked on simultaneously, and he soon found that being employed by Western Union took valuable time away from inventing. With little money of his own, in January 1869 he nevertheless quit his job, cleared some space in the workshop of a Boston telegraph maker, and set to work, determined to make his fortune. He found enough financial backing to enable him to develop an improved stock ticker, for which he filed his second patent; and he found about thirty companies willing to rent his models. This success allowed him to hire a few workers, and to manufacture and set up his own telegraph system, stringing his lines alongside those of Western Union. "It never occurred to me to ask permission from the owners," he admitted, "all we did was to go to the store, etc. and say we were telegraph men and wanted to go up to the wires on the roof and permission was always given."[6] Besides the stock ticker, he worked tirelessly to solve the problem of sending two messages over a single wire, an achievement that would substantially speed up telecommunications. With an $800 loan he constructed this duplex telegraph, and with much excitement pressed for a chance to demonstrate it to Western Union. When the company refused, he immediately took the idea to one of its rivals, the Atlantic and Pacific Telegraph Company in Rochester, and he set out for western New York eager to make a sale; but when he tried to communicate between Rochester and New York City, the telegraph unaccountably failed. Edison had no choice but to return to Boston; he was left practically destitute.

If other men felt discouragement at such setbacks, Edison did not. By June 1869 he was in New York City, hoping to find a job at Western Union so that he could shore up some savings and return to inventing. While he waited to begin, he came to the rescue of the Gold Indicator Company, whose instruments had broken down suddenly. Within a few hours, Edison located the problem and reset the devices. Since the company's electrical engineer, Franklin L. Pope, had just resigned to start his own business, Edison was hired immediately to oversee the plant at a salary of $300 a month, "such a violent jump" from what he had ever earned that he vowed to work twenty hours a day if necessary to prove himself worthy of it.

Being a site supervisor, however, was not one of Edison's career goals. In October, he and Franklin Pope formed Pope,

Edison and Company, consultants and inventors. They found a space for a workshop in Jersey City, where Edison spent some seventeen hours a day. Among other projects, he and Pope developed a telegraph system to compete with Western Union at much lower prices, and their subscribers increased at such a rate that within six months, Western Union bought out the company for a sum larger than anything Edison had ever imagined would be his: $15,000. "I have too sanguine a temperament to keep money in solitary confinement," he remembered, "so I commenced to buy machinery, rented a shop," hired men to assist him, and worked day and night, with only a few hours of sleep occasionally interrupting his manufacturing and inventing.[7] Among the machines that he and his "muckers," as he called his staff, worked on were a Regulating Temperature Machine, perforating machines, a screw-slotting machine, a wire-straightening machine, printing telegraph instruments, galvanometers, fire alarms, and improved versions of anything he or others already had invented.

In 1869 Edison was twenty-two, and through a combination of ferocious ambition, talent, and arrogance, had positioned himself at the forefront of his field and of an electrical revolution. But that revolution had not yet made an impact into everyday life. The few battery-driven devices that had made their way into the home proved so unreliable and frustrating that few people saw the benefit of owning them. Burglar alarms required continuous cleaning to keep dust and moisture out of the battery. Electric call bells, used to summon servants in some homes and as doorbells in others, frequently broke down because their battery wore out or became clogged with dust. Even when working as they should, batteries lasted only a matter of weeks before they needed replacement, making any electrical device an expensive item to maintain. Although Edison tossed one invention after another into the marketplace, that marketplace was limited to businesses.

THE CHIEF

If Edison had seen the home as a viable market, he might have devoted some of his efforts to inventing devices for household use. But from the time that he sequestered himself in his laboratory, away from his family, Edison had little experience of domestic life. Home was a place where he slept occasionally, but often it was more convenient simply to grab a few hours' nap in the corner of a workshop or stretched out on a laboratory table with a few books for a pillow. Edison seemed oblivious to the need for refuge or relaxation or, oddly, love. It came as a surprise to many who knew him when he discovered the charms of sixteen-year-old Mary Stilwell, who worked as a telegraph tape perforator at one of Edison's short-lived enterprises, the News Reporting Telegraph Company. Mary was attractive and winsome. There was hardly a courtship: Edison took her on a few carriage rides before he asked her father for her hand in marriage. Stilwell, reluctant because of Mary's age, asked Edison to wait a year before the wedding. If Edison agreed, he did not keep his word: the couple were married three months later, on Christmas day of 1871. That evening, he was back in his workshop.

They did take a honeymoon trip, accompanied by Mary's older sister, who then moved in with them in Newark. The family grew to five, with the birth of Marion in February 1873, whom Edison nicknamed Dot; Thomas Jr. in January 1876, called Dash; and William Leslie in October 1878. The two oldest children's nicknames were not really whimsical. Edison's most cherished progeny, as the family discovered, were his inventions. With Edison working long hours, often not returning home for days at a time, Mary—Popsy Wopsy to her husband—and the children engaged in their own social and family life. Edison had little idea of what Mary did all day, what she and her servants needed to run a household, or how the children spent their time. His only basis for imagining a child's life was his own youth. In 1874, he proposed to set up a Scientific Toy Company, which would manufacture twenty-eight items for children's use, including a Magneto-electric-shocking Machine, a microscope, an Air Pump, and a Turbine Wheel. Although a loom, fire engine, "Sewing Machine for Little Girls," and electric boat also were offered, the list is weighted with items that Edison himself might have dreamed of owning as a child, items to furnish a cellar workshop but not a playroom.

Edison did understand, however, that in the world of business—including his own—time was money. For much of the 1870s he staked his reputation on speed: increasing the speed of telegraph communication and rushing to file caveats—protection for an idea while it was being developed—and patents more quickly than any competing inventor. Although the telegraph was touted as the annihilator of time and was established as a vital part of business life, Edison knew that the machine still worked too slowly. As the number of messages sent increased, the speed with which they arrived at their destination slackened. In some parts of New York City, it was quicker to send a messenger on foot than to wait for a communication over the wires. Edison saw two potential improvements for the telegraph's efficiency: to enable messages to be transmitted automatically from one station to another and to send several messages simultaneously.

Because wires did not connect every telegraph office directly to every other, a message often traveled from one office to a hub, where it would be routed to another office and even to a third, with operators needed at every stage of transmission. But if the message could be transmitted directly onto paper as indentations or perforations, Edison thought it would be possible to lessen the need for operators as well as to speed transmission. Once he developed this automatic telegraph, he set off to market it, this time beginning in England, and in the spring of 1873, he embarked on his first trip abroad. His six-week journey was frustrating: demonstrations did not work as well as he hoped, and Edison discovered that he had competition from British patent holders for similar devices. After he returned, when Western Union also refused to adopt the automatic telegraph, he took an offer from Jay Gould, the railroad magnate and owner of the Atlantic and Pacific Telegraph Company.

Edison often had thorny relationships with partners and investors if he thought they threatened his authority, but his relationship with Gould was particularly rancorous. For one thing, Gould, Edison noted, apparently had no sense of humor. Edison, used to getting attention and admiration by telling amusing stories, found himself unable to establish any kind of rapport with Gould, who was interested in only one thing: making money. And his power, Edison discovered, was the result of obsession and ruthlessness. "I think Gould's success was due to abnormal development," Edison wrote later. "He certainly had one trait that all men must have who want to succeed. He collected every kind of information and statistics about his schemes and had all the data... His conscience appeared to be atrophied," Edison added, "but that may be due to the fact that he was contending with men that were worse."[8] Despite Gould's lack of scruples, Edison had no problem taking money from him nor in siding with him against Western Union when he went after control of Edison's former employer.

In the end, Gould jilted Edison and his backers, and their business dealings ended in the mid-1870s. However indignant Edison felt about Gould's tactics, his own behavior with competitors suggests that he learned something about intimidation from Gould.

Besides speeding up message routing, Edison continued to develop devices able to send more than one message along the same wire at the same time: duplex, quadruplex, and even octuplex telegraphs. The pace of daily office work offered another problem for him to solve: how to speed up the duplication of documents, a time-consuming activity that involved repeated copying. His solution was the Edison Electric Pen, a battery-powered stylus that perforated a sheet of paper as one wrote. The tiny perforations then were filled in with ink from a felt-covered roller and pressed against a sheet of paper held steady against a metal plate, making it possible to duplicate a document any number of times. To Edison's great satisfaction, the electric pen proved so popular that within a few years it became an essential business machine; Edison sold British and Canadian rights, and hired agents to market it in Europe and South America. Eventually he sold rights to the invention to A. B. Dick, who improved the device and sold it as the Edison Mimeograph.

By late 1875, Edison felt that he had outgrown his rented Newark facilities and envisioned a workshop and laboratory that would give him enough space and equipment to experiment, develop and test ideas, manufacture machinery, and meet the orders that he hoped would result from his inventions. At the end of December, he bought two tracts of land and a house in the tiny hamlet of Menlo Park, twelve miles south of Newark. Three months later, he was ready to move into his newly completed industrial park, the largest private laboratory in America.

At one hundred by thirty feet, the two-story structure, with its white clapboard exterior, front porch, and white picket fence, looked more like sleepy town offices than a factory and research facility fitted with the most up-to-date electrical equipment, generators, measuring instruments, and over two thousand bottles of chemicals. More than a hundred muckers could work comfortably in the ample spaces. Edison moved Mary and their two children into a three-story house nearby—it, too, surrounded by a white picket fence. But it was hardly the house of Mary's dreams. One of nine other dwellings in Menlo Park, this house was so far from the others that Mary took to sleeping with a revolver under her pillow on the nights, and they were many, when her husband did not come home. Her fear was so great that soon Edison bought three dogs to protect her.

Edison was much more comfortable in the gritty atmosphere of his workshop than in the home that Mary had decorated with lace coverlets and curios. At work, Edison was the leader of a band of loyal men—"our chief," they called him, or "the old man"—who respected his ingenuity and tenacity, humored him for his practical jokes and the homeopathic concoctions that he mixed up whenever a worker complained of any ailment, and learned to keep out his way during his violent outbursts. Edison's temperament, one associate recalled, was "essentially mercurial," ranging from exuberant high spirits at one moment to dark depression or rage the next. If work went well, everyone basked in his pleasure. If he became angry with one worker for some mistake or another, all the others would suffer, unless they managed to avoid him.

One characterization of the muckers came from George Bernard Shaw, who worked in Edison's British telephone company for a few months in 1879. His co-workers, Americans who had been trained by Edison, were, Shaw recalled, "deluded and romantic men [who] gave me a glimpse of the skilled proletariat of the United States. They sang obsolete sentimental songs with genuine emotion; and their language was frightful even to an Irishman. They worked with a ferocious energy which was out of all proportion to the actual result achieved." They refused to take orders from an Englishman, and instead "insisted on being slave-driven with genuine American oaths for a genuine free and equal American foreman. They utterly despised the artfully slow British workman who did as little for his wages as he possibly could; never hurried himself; and had a deep reverence for anyone whose pocket could be tapped by respectful behavior." The British, at the same time, wondered why they worked so hard to put money in someone else's pocket. The American workers "adored Mr. Edison as the greatest man of all time in every possible department of science, art and philosophy... They were free-souled creatures, excellent company: sensitive, cheerful, and profane; liars, braggarts, and hustlers; with an air of making slow old England hum which never left them."[9] This was the excellent, irresistible company that kept Edison in his laboratory rather than in Mary's parlor.

Edison's attention sometimes was diverted from his work, however, by legal entanglements. The early days at Menlo Park were blighted by the many suits in which Edison managed to become involved, or to initiate, throughout his career. Nevertheless, soon Edison returned to inventing, this time to improving what he called the speaking telegraph and Alexander Graham Bell called the telephone.

Bell's interest in reproducing human speech had begun when he was sixteen and tried to construct a speaking automaton, worthy of the eighteenth-century French master Jacques Vaucanson, with jaws, teeth, and nasal cavities molded from gutta-percha; a flexible, segmented wooden tongue; and fleshy rubber lips. Even Bell thought that the "speaking machine," features without a face, looked truly horrifying, yet he managed to get some faintly human squawks to emerge from the mouth.[10] Thirteen years later his goal was far different: to devise a machine that could reproduce the speech of one individual and convey that speech, over a distance, to a listener whose reply would be reproduced in turn. Throughout the early 1870s, Bell worked on various models and finally, in May 1876, felt confident enough to demonstrate his telephone before MIT scientists at the Boston Athenaeum. That day the model worked well, and the usually staid audience broke into astonished applause at hearing a voice that originated twenty miles away. Buoyed by that success, a month later Bell brought a model to the Centennial Exhibition that recently had opened in Philadelphia. As in Boston, fellow scientists and inventors admired his achievement, but surprisingly there was little response from the press. While Edison won awards at the exhibition for his automatic telegraph system and electric pen, Bell left as unheralded as he had come.

During the summer and fall of 1876, Bell continued to work on his models, focusing especially on two persistent problems: clarity and volume. With few exceptions during trials, the speaker had to shout repeatedly to be heard, and sibilant sounds—*s*, *th*, and *sh*—were indistinguishable. The telephone was far from perfected, and Bell knew that many other inventors were working to produce a better instrument. It seemed prudent to sell rights for his own version before those competitors invaded his territory. But when one of Bell's partners approached Western Union late in 1876, the company turned him down: with the telegraph becoming increasingly efficient, thanks to Edison, what would be the use of communicating by voice? Although the renowned scientist Sir William Thomson, later elevated to Lord Kelvin, told Bell that the telephone was "the most wonderful thing" he had seen at the Centennial Exhibition and predicted that "before long, friends will whisper their secrets over the electric wire," this buzzing gossip did not seem to portend an adequate market.[11]

Yet at the same time that Western Union spurned Bell, it encouraged Edison to see if he could make a superior instrument, and make it quickly. Edison experimented with various configurations of a vibrating tube and various kinds of material for the diaphragm, finally discovering that carbon, much more sensitive to sound waves than Bell's hammered metal diaphragm, offered the answer to both the clarity and volume problems that Bell had encountered. "Perfection," Edison announced, occurred at 5 A.M. on July 17, 1877. Within months, Western Union organized the American Speaking Telephone Company to compete with Bell, who had started his own company in Boston. As Edison put it, "the fight was on [with] the W.U. pirating the Bell receiver and the Boston Co. pirating the W.U. Transmitter." The fight included competing claims for patent rights, which finally were awarded to Bell in the fall of 1879. Western Union withdrew from the telephone business in exchange for a royalty agreement, and Edison, who told Western Union that he wanted some compensation, agreed to a payment of $100,000 to be distributed over the seventeen-year life of his patent.[12] Still, it was unclear what use the telephone would have. For businesses, the telegraph served the purpose of conveying information or instructions, and even when messages required a reply, the interaction could not be construed as a conversation. The telephone, providing an opportunity for two people to interact when they were apart physically in the same way they would in the same space, failed to capture the public's imagination. The definition of *call* was *personal visit.*

In fact, some people believed that the telephone would be used just as a telegraph, with operators forwarding voice messages, receiving them back, and relaying them to the customer.[13] Certainly few saw a market for telephones in the home, except perhaps as a kind of intercom to summon servants.

Inventing a need for this new device proved to be a formidable challenge. One journal of applied science speculated that the telephone might have several "novel" uses, such as providing a means of communication between vessels at sea; or allowing a policeman to contact his stationhouse in an emergency with a portable telephone that he could attach quickly to wires in a special call box; or enabling a clergyman, too far from his parish on Sunday morning, to call in a sermon from wherever he was.[14] In businesses where information needed to be sent continually between one office and another—an order department and a factory, for example, or a publisher and printer—the telephone might be more efficient than a telegraph. And in communities too small to support a telegraph office, a telephone line might be used to send messages to the nearest telegraph office. But a telephone in the home seemed frivolous and potentially invasive. Even Bell saw home use only in the service of consumption; he thought that homemakers might want to phone their orders for food or household goods. By the time some fifty thousand telephones finally were installed, most were found in offices.

CAPTIVITY OF SOUND

As Edison worked on improving the telephone diaphragm, he diverted himself by making a little mechanical toy. This toy, though, was powered not by a battery but by sound. When Edison spoke into a funnel at the top, sound waves caused a diaphragm to vibrate, engaging a gear that moved a pulley connected to a small paper figure. "Hence," he said, "if one shouted Mary had a little lamb, etc., the paper man would start sawing wood." The toy convinced Edison that the diaphragm could solve a problem he had been working on for some time: how to record telegraphic, and eventually telephonic, messages to provide a permanent record for businesses. One way was to transfer telegraph messages as perforations on tape; another, clearly, was to record and reproduce the human voice. Throughout the summer, Edison worked on variations of this idea. "Do you know Batch," he told Charles Batchelor, his chief assistant, "I believe if we put a point on the centre of that diaphragm and talked to it whilst we pulled some of that waxed paper under it so that we could indent it, it would give us back talking when we pulled the paper through a second time."[15] By early November, after he substituted foil wrapped around a cylinder for strips of waxed paper, he was certain enough of the idea that he allowed Edward Johnson, another of his staff, to announce the invention as a letter to the editor of *Scientific American*. Although Johnson testified that the apparatus was still "crude," he said that Edison planned to have the phonograph "in practical operation within a year." In fact, Edison had a model ready within weeks, and on December 7, he walked into the offices of *Scientific American* carrying his newly invented "little machine." He turned a crank and, reported the dozen or so staff who witnessed this miracle, "the machine inquired as to our health, asked how we liked the phonograph, informed us that *it* was very well, and bid us a cordial good night." The phonograph spoke clearly and audibly, the journal reported later in the month, and it astonished everyone who heard it.[16] The journal's staff insisted that Edison play the machine over and over, and as the crowd of listeners grew to include local reporters, the editor became afraid that the floor would collapse. Although Edison tried to explain clearly the principle behind the phonograph, reporters were incredulous. "It was so very simple," he said, "but the results were so surprising they made up their minds probably that they never would understand it—and they didn't."[17] Newspapers soon carried accounts of the latest technological miracle. Within days of the demonstration, Edison filed a patent.

Yet just as with the recently invented telephone, which in 1877 had made no significant impact in home life, the phonograph seemed like nothing more than a new toy. Edison's technical note to himself at the end of November suggests that even he was struggling to think of what the phonograph might do besides serve businesses. Basically, he saw it as a personal recording device, much like a camera. To make it more marketable, he strained to think of a larger number of occasions for which people might welcome sound. His deafness, in this effort, was not an asset. "I propose to apply the phonograph principle to make Dolls speak sing cry & make various sounds also to apply it to all kinds of Toys such as Dogs' animals, fowls, reptiles, human figures: to cause them to make various sounds to steam Toy Engine imitation of exhaust & whistele [*sic*]." Beyond toys, he imagined that the phonograph could reproduce vocal or orchestral music for a family's "endless amusement" and for music boxes. One original idea was for a speaking clock that would call out the time of day and a friendly reminder of one's appointments. Phonographs could bark out advertisements; in fact, Edison was working on amplification so that these exhortations could be heard. With such tenuous ideas about its usefulness, it is no wonder that Edison could not induce Western Union to back the development of the phonograph.

By the time "The Phonograph and Its Future," ghostwritten by Edison's assistant Edward Johnson, appeared in the *North American Review* six months later, Edison was able to make a stronger case, conceding, nevertheless, that "the public press and the world of science" might contribute some "imaginative work of pointing and commenting upon the possible."[18] One impediment to his own imagination seems to have been the difference between this invention and his previous devices: this time, rather than selling speed, Edison extolled the marketability of permanence. The phonograph, he explained, provided for the "captivity of all manner of sound-waves heretofore designated as 'fugitive,' and their permanent retention." The reproduction, true and faithful to the actual source, could be played back from fifty to a hundred times—enough repetition, he thought, "for all practical purposes." One could record a message to a friend, for example, remove it from the phonograph and place it in a special envelope, send it by mail, and allow the recipient a chance to hear one's voice over and over—provided, of course, that the recipient owned a phonograph. But this social nicety seemed less important to Edison than its potential advantage in business: at last he had found a way to give businesses a permanent record of communications. Moreover, recording letters directly meant dispensing with clerks or stenographers, not only saving money but insuring "perfect privacy" and accuracy in communications. Even if clerks were kept on the payroll, managers could use the phonograph as a dictating machine, making it possible for transcription to occur at flexible times.

Owners of phonographs possessed power: the phonograph, after all, afforded "perfect privacy" only to those who operated it. Recordings, Edison suggested, could be made without "the knowledge or consent of the source of their origin." This "essential feature" of the phonograph implies that it could be used surreptitiously at conferences or business meetings, turning private conversations into public record or underhanded means of manipulation.

Edison repeated his ideas for talking toys, clocks, and advertisements, and added some new uses, perhaps suggested by his co-workers or audience members at his demonstrations: books read onto disks for the benefit of the busy, the ill, or the blind; recording of testimony in court, resulting in "an unimpeachable record"; educational lessons in the form of correctly spelled words, facts to be memorized, and foreign languages to be drilled. Most significantly, it would provide a permanent archive of personal or collective memory. The speeches of great leaders could be heard forever, or at least as long as a disk lasted, and the phonograph could preserve "the sayings, the voices, and *the last words* of the dying member of the family." If the telegraph annihilated time, the phonograph made it possible to attenuate, relive, and preserve the present.

The popular press taught the public how to receive the new inventions of the telephone and phonograph, inventions to which most people had no access except at exhibitions. Reporters described these instruments as sources of amusement and entertainment, inventions that anyone would want to see, just as an electrified model insect or an electrified "Venus" had drawn eighteenth-century audiences. The phonograph was just the latest of these curiosities, and newspaper and magazine writers testified to being enchanted by the "mysterious" machine that worked "in some inexplicable way" to reproduce sound. "The other day it was the telephone that was filling every one with amazement," commented an editorial in *The Atlantic Monthly*, "but that promises soon to be forgotten, or at least to lose its novelty, by the side of the greater wonders of the phonograph." Those wonders included melodramatic scenarios in which hidden phonographs recorded "the whispered plottings of conspirators" or "the soliloquies of villains."[19] The phonograph, stealthily, might detect and reveal unexpurgated truth.

Those who actually used the instrument in the workplace, where some businesses decided to try the phonograph as a recording device, found that it could not deliver on Edison's promise of permanence. The machines were fragile and unreliable, so sensitive to dirt and vibrations that they frequently stopped running. To make a recording, a crank needed to be turned at a precise speed as the speaker enunciated into a mouthpiece; when the recording was played back, the phonograph needed to be turned at exactly the same speed or the results were distorted, a bass voice emerging as a soprano, a soprano rising to a piercing squeak. Even when trained technicians made a recording, the tone seemed shrill. The foil disks did not last even a few replays, much less the fifty to a hundred that Edison predicted, and no one knew how to send them through the mail without damaging them. Edison had considerable work to do, but in 1878 the phonograph seemed unlikely to find a large market. "The earning capacity of the phonograph," two contemporaries observed, "lay in its exhibition qualities."[20] Rather than devote more time to the intensive work necessary to perfect the instrument, Edison moved into an entirely new and potentially more lucrative field of research: illumination.

CHAPTER 3

Wilderness of Wires

Edison's demonstration lamp, 1879. The hand-blown bulb is darkened by carbon evaporated from the carbonized bristol-board filament.
The Smithsonian Institution

I felt the sense of great responsibility, for unknown things might happen on turning a mighty power loose under the streets and in the buildings on lower New York. However, I kept my counsel.

Thomas Edison

On September 8, 1878, Thomas Edison visited the brass and copper foundry of William Wallace, in Ansonia, Connecticut, to look at a new electric generator—the "telemachon," Wallace called it—that drew its power from a river a quarter of a mile away rather than from a battery or steam engine on site. What Edison saw that day astounded him: eight electric arc lamps, emitting 4,000 candlepower, blazed at once, energized by the telemachon: in short, Wallace had achieved a rudimentary lighting system. According to a reporter who accompanied him, Edison "was enraptured," running from the instruments to the lights and back again, calculating the power used, the power lost, and the cost of transmission for "a day, a week, a month, a year." Immediately, Edison saw the economic potential and recognized the next step: to subdivide the lighting so that each lamp would have an intensity suitable for interior use, and to create a dynamo that could power even more lamps. Wallace's experiment showed him a model; now, he had to perfect it.[1] And he had to perfect it quickly.

By 1878, electric lighting existed in the form of arc lamps, which emit a startlingly intense bluish beam. In an arc lamp, illumination occurs between the tips of two carbon rods laid end to end. For the light to be steady, the gap between the rods needs to be regulated so that it remains the same while the carbon burns; this regulation was not easily achieved. Besides the gap problem, arc lamps often flickered, sputtered, and hissed, the way a log does burning in a fireplace. The sputtering and the unsteadiness, along with their intense brightness, made arc lamps unsuitable for small interior spaces. And, some believed, they were not suitable for large outdoor spaces either. Because of its intensity, the arc lamp needed to be set on a high pole, well above eye level; because the lamp distributed illumination at a forty-five-degree angle, the effect was like a spotlight: bright pools of light were surrounded by darkness. In some cities, several arc lights were mounted on a tower, dispersing brilliant illumination for a circumference of hundreds of feet and leaving the area beyond in total blackness. The effect was sensational to some, eerie to others, glaring for all.

Objections to arc lighting echoed earlier objections to even the idea of public lighting. In 1816, the installation of gaslights in Cologne seemed an abomination: artificial illumination, a newspaper article protested, "is an attempt to interfere with the divine plan of the world, which has preordained darkness during the night-time." Rather than deter crime, public lighting would make it easier for thieves to work; and without the "fear of darkness . . . drunkenness and depravity [will] increase."[2] The new arc lighting, already installed at some London and Paris sites, appeared worse; Robert Louis Stevenson called it "horrible, unearthly, obnoxious to the human eye; a lamp for a nightmare!" Or an insane asylum, where it would be "a horror to heighten horror. To look at it only once is to fall in love with gas, which gives a warm domestic radiance fit to eat by." Humans were misguided to explore "the profound heaven with kites to catch and domesticate the wildfire of the storm," he wrote. Civilization should be carried out beneath "the old mild lustre" of gas lamps.[3]

But besides emitting a mild luster, gas lamps also gave off heat, soot, and noxious fumes. A theater lit by hundreds of gas sconces quickly became stifling; in every home, furniture and walls darkened over the years; and a gas leak in a nursery was a mortal threat. Moreover, any home lit by gas was at the mercy of the companies that supplied it and, many consumers complained, their greedy owners. Still, gas was familiar, and certainly an improvement over oil or paraffin lamps, which required daily cleaning and frequent attention. Among the wealthy, affection for gas tended to increase in relation to one's stock holdings. Economic concerns, as well as aesthetic ones, might well have inspired the protest verse that appeared in the *St. James Gazette* after arc lighting was introduced at Paddington Station in London.

Twinkle, twinkle little arc,
Sickly, blue uncertain spark;
Up above my head you swing,
Ugly, strange expensive thing!

Cold, unlovely, blinding star,
I've no notion what you are,
How your wondrous 'system' works,
Who controls its jumps and jerks.[4]

Nevertheless, arc lighting became the focus of research for several inventors—fueled by the spark of inventiveness and hope of a fortune—who were certain that lighting by electricity would prove cheaper for cities and towns than lighting by gas. Moses Farmer, a collaborator on Wallace's telemachon dynamo, had exhibited arc lights at the Philadelphia Exposition in 1876; Russian electrician Paul Jablochkoff devised an improved arc light, known as the "Jablochkoff candle," in which the carbons stood upright, parallel to each other, solving the problem of regulating the space between the carbon tips; Joseph Swan was working assiduously in England, as was Werner von Siemens in Berlin. But Edison's chief American rival was the tall, gregarious twenty-eight-year-old Charles Francis Brush.

Like Edison, Charles Brush, even as a boy growing up in Euclid, Ohio, was an inventor. At Cleveland's Central High School, he began to experiment with electricity, generating static electricity, and constructing Leyden jars, batteries, electromagnets, and motors. He was particularly fascinated with the arc lamp, and after many failures, finally achieved a success. He was sixteen, and, he recalled, "it filled me with joy unspeakable."[5]

After graduating from the University of Michigan in 1869, Brush returned to Cleveland and worked as an analytical chemist, devoting all of his spare time to inventing, especially to developing an economically efficient dynamo and to improving the arc lamp. During the summer of 1876, holed up at his family's farm, he focused only on those two projects. For the lamp he contrived a mechanism that could adjust the carbons as they burned, as well as a device that restarted combustion if the current were interrupted. And he experimented with several models of dynamos until he found one that he believed would produce sufficient current to power his strong lamps. He started the dynamo first with a single-cell battery, then, after more hopeful trials, took a model to the workshop of the Telegraph Supply Company and tested it using steam power. The president of the company, a friend of Brush's, agreed to finance the manufacture of the dynamo, and by the summer of 1877, at Philadelphia's Franklin Institute, Brush proudly exhibited his own arc lights powered by his new dynamo.

The demonstration inspired interest from Elihu Thomson—Edison's former adversary, with whom Brush formed a professional relationship—and from a Cincinnati physician, who bought the first dynamo and lamp to be exhibited publicly. The effect was more than dazzling: the 4,000-candlepower light was, without doubt, the brightest artificial illumination that anyone had ever seen. Among the large crowd gathered beneath the doctor's balcony, many, knowing nothing about electricity, assumed that the new lamp was an innovation in oil lighting.

In the next few months, Brush received commissions from businesses, including the cavernous John Wanamaker department store and a four-story carpet factory, both in Philadelphia. As Brush marketed his products, he realized that he needed to educate the public about how to avoid "light blindness." "The principal difficulty arose," he explained, "from the propensity of everybody to stare directly at the arc, and then declare that everything else looked dark. It took years fully to outgrow this habit." Another problem arose when those in charge of maintaining the new systems tried to adjust the lamps themselves, resulting in lost screws or other parts. Brush tried to invent a tamper-proof design, but never managed to make one completely foolproof: lamp users, he discovered, were craftily ingenious.[6]

By the time Edison saw Wallace's telemachon dynamo, Brush had installed a six-light series in a Boston clothing store and was able to offer sixteen-light packages—the largest his dynamo could support—which he hoped would interest owners of factories, theaters, and hotels. Even Brush's improvements, however, did not make arc lamps ideal: if the lights stayed on for a few hours, the carbon needed to be replaced several times. As Brush worked to further improve his lamps, Edison hurried to corner the field. A week after visiting Wallace's shop, he announced that he had successfully subdivided light. "I have it now," he informed a reporter for the New York *Sun*, "and, singularly enough," he added, implying some secret magic, " I have obtained it through an entirely different process than that from which scientific men have ever sought to secure it." That "different process" depended on inventing lights with a high resistance to electrical current; a system using low current would require less expensive copper wire to connect lights to their source of electrical power and would enable a generator to power more lights. He could produce a thousand, maybe ten thousand, lights from one generator, he said; and with fifteen or twenty dynamos, he was certain that he could light the entire business area of lower Manhattan.[7] Edison rushed to publicize his discovery less to ward off competitors than to attract investors. Within a month, the Edison Electric Company was incorporated, backed by $300,000; when word of the new company reached England, it set off a panic in London gas shares. American investors also felt fearful, although some important holders of gas stock decided to protect themselves by investing in Edison's new project. Among Edison's backers, J. P. Morgan had faith that the new technology would "prove most important . . . to the world at large [and] to us in particular in a pecuniary point of view. Secrecy at the moment is so essential that I do not dare to put it on paper. Subject is Edison's electric light."[8]

Meanwhile Edison experimented with his lamps, which were proving more frustrating than he had anticipated. If he believed that he knew, in theory, how to subdivide light, still he did not have functioning lamps. Still, he faced problems that had beset inventors for forty years. The lamp required a delicate interaction of a metal that could be heated to give off a steady glow—the meaning, after all, of incandescence—a regulator to prevent the metal from becoming overheated and quickly consumed, and a durable burner to hold the bulb. As early as 1838, developers of incandescent light recognized the importance of creating a vacuum within the bulb, but that goal seemed elusive to Edison's staff as they worked to develop an improved vacuum pump. Then there was the matter of the bulb itself. Each was handblown, and though Menlo Park could keep up its supply for experimentation, Edison needed to develop a way to produce bulbs cheaply and in huge quantities.

The greatest challenge, the problem that his predecessors had found insurmountable, was the filament. Within the delicate glass globe that filament needed to glow without consuming itself, to provide illumination no more glaring than a gas burner, to emit light without also emitting gases that would darken the globe, and to be inexpensive enough to make a lighting system affordable. Within weeks of his hyperbolic announcement to *The Sun*, Edison filed several caveats for ideas that he thought would work, but none did. Others before him had tried carbon and platinum, but each had its disadvantage: carbon burned only for eight minutes; platinum, glowing at an exceedingly high temperature, melted within ten. Nickel, cheaper than platinum and for a while Edison's great hope, oxidized too quickly.

Publicly, Edison sounded optimistic. Privately, he was not. "I have only correct principle," he wrote to one of his European representatives early in October, referring to his plan for high-resistance lights. "Requires six months to work up details."[9] As month after month passed, however, reporters and Edison's backers grew impatient. Rumors that other inventors had succeeded in producing an incandescent bulb caused some investors to press for assurances of Edison's progress. Visiting Menlo Park somewhat assuaged their worries; they also had to remind themselves of Edison's record of success. The inventor who had given them the quadruplex telegraph, the electric pen, and the improved telephone would, they believed, give them a lighting industry.

But problems multiplied, and in the spring of 1879, the Menlo Park staff devoted themselves to identifying a successful filament, devising a way to evacuate the bulb as completely as possible, inventing a way to measure usage, and designing a dynamo more powerful than Wallace's. At the same time, Moses Farmer, in a published interview, raised other problems that Edison seemed not to have considered: before an electrical system could be established, there would have to be a sufficient number of electrical engineers able to support such a system, Farmer cautioned, and telegraph electricians simply were not prepared to do so; laying copper wires underground meant that a proper insulating material needed to be found, which at the moment was uncertain; and, of course, the public would have to be enticed to switch from gas lighting to electric lighting.[10]

Electric lighting in London provided evidence for Farmer's cautions: at the Holborn viaduct and the Post Office, all of the arc lights sometimes were extinguished completely because if one lamp failed, all would fail. "The experiment," the journal *Nature* commented, "seems to be conducted by some one who is not experienced in the working of electric circuits." Whenever there was mist in the air, or even high humidity, the lights might go out or, even if they did not, could hardly be seen. Unless durability and steadiness could be ensured, the journal concluded, "[t]he present state of the electric light question may therefore be said to be a tentative one."[11]

Brush confronted some of these problems when he installed lighting systems in mills in Providence, Rhode Island; Hartford, Connecticut; and Lowell, Massachusetts; and in a San Francisco hotel and several dry-goods companies in New York. Finding technicians who could operate the systems, as Farmer had predicted, proved a huge problem. "Once I traveled fifteen hundred miles to take a common staple tack from the bottom of a dynamo, where it happened to short-circuit a field-magnet," Brush recalled.[12]

Lack of technical support, however, did not deter a few cities from buying arc lights in the hope that they would be more economical than gas. In April 1879, Cleveland became the first American city to buy a Brush lighting system; twelve arc lamps around Monumental Park transformed it into a center of entertainment. When the moment came to turn on the lights, Brush saw that many of the thousands gathered for the festive event had prepared themselves for illumination as bright as the sun's glare by wearing colored glasses or holding up smoked glass. A local band began to play, and from the lakeshore an artillery squadron fired a celebratory salute. As the onlookers watched in silence, each lamp blazed in succession until the whole square was lit. The effect was not what anyone expected: people's skin turned a bluish gray, the colors of their clothing looked faded and dull. "Of course," Brush noted, "there was at first a general feeling of disappointment in this respect, although everyone was willing to admit that he could read with ease in any part of the square."[13] Or not quite with ease: the lamps, everyone soon discovered, produced an unsteady glow, flaring unexpectedly into painful brightness. Still, seeing the park newly illuminated was a harmless diversion, much as attending a demonstration by Davy or Faraday had been for a previous generation. If public lighting had a usefulness beyond possible cost saving, it eluded most onlookers.

It took a year before another city decided to order a Brush system: Wabash, Indiana, faced with the expensive cost of gas for lighting its downtown, asked Brush to provide a single lamp that would illuminate a half-mile radius, or most of the city, for which he would be paid $100 for a month's trial. If the system met with approval—that is, if it was cheaper than gaslight—the town would then proceed to buy it. That plan, cautious though it seems, generated bitter criticism from many who felt the town was wasting its money. "Why not make a contract with the man in the moon?" a prominent newspaper editor asked. "He'll furnish light half the time anyway."

Some townspeople's concerns focused on agriculture. If the lights turned night into day, one resident worried, "and as chickens never sleep during daylight it is only a matter of time when every fowl within the corporate limits of Wabash will die for lack of sleep." Perpetual light on cornfields might mean that corn would grow so huge it would have to be cut down with saws. Arc lamps, for farmers, might prove an expensive folly. For most people, though, the coming of light meant nothing other than an evening's excitement. Notices of special excursion trains to see the illumination assured visitors that they would gaze upon a "marvel of the nineteenth century," "a light that shows all the beautiful colors as distinctly as the sun, and gleams as pure and white as the full moon." So much publicity preceded the first lighting that the spectacle might have seemed anticlimactic. But at eight o'clock on March 8, 1880, when Brush's four arc lamps were finally lit, the thousands milling in the town center fell into a stunned silence. "The people, almost with bated breath, stood overwhelmed with awe," said one eyewitness, "as if they were near a supernatural presence." One elderly man living on the outskirts of town had not heard of the event; when light flooded his barnyard, he ran into his house, trembling with alarm. "Down on your knees, Mary!" he cried out. "The end of the world's here!"[14]

New York City was Brush's next commission; he incorporated the Brush Electric Light Company in September, and by mid-December installed generators at West Twenty-fifth Street to light three-quarters of a mile of Broadway with fifteen lamps, each set on twenty-foot poles. At 5:30 P.M. on December 21, the current was turned on, and, as a *New York Times* reporter described it, "The white dots broke out—not one after another, but instantaneously, as though a long train of powder had been fired." Christmas shoppers looked up from gaslit shop windows to admire the "artistic effects of strong contrasts of light and shade"; in front of Tiffany's an elegant carriage drawn by two white horses, lit brilliantly against the darkness, "formed a picture." Passersby commented, hopefully, that the gas monopolies soon would be out of business.[15] Compared with the flickering gaslights that now were extinguished, these new lights were resplendent. But as New Yorkers soon discovered, arc lighting, picturesque though it might be, was unreliable, now and again failing altogether and leaving Broadway in total darkness. Technicians struggled to find a cause, sometimes blaming wet weather, sometimes a break in the machinery; and sometimes, they told reporters, they were "too busy" to answer questions.[16] Within a few months, frequent failures made the viability of arc lights on Broadway uncertain, a development that pleased Edison and his backers. As Brush set out his wires and mounted his lights, they had incorporated the Edison Illuminating Company of New York City.[17]

PROFESSOR OF DUPLICITY

His reputation, Edison discovered as he worked on his lighting project, was fragile. Although the press referred to him, with affection, as the "Wizard of Menlo Park," he realized that he was only as famous as his last success; and as his business dealings became increasingly public, his image as guileless wizard, a recluse in a wilderness of wires, seemed hard to believe. Edison's repeated claims about his progress, claims that were not supported by successful demonstrations, frustrated reporters, and they began to suggest that he might be an accomplice in the panic in gas shares that resulted from each announcement. Could it be that Edison was a ruthless businessman?

In the fall of 1879, *The Sun* wondered why no light had appeared in the year since Edison had promised to illuminate Menlo Park. "Edison has said over and over during the year that he had solved the problem of the 'indefinite subdivision of electric light.' Far from this being true, he has not even approximated a solution, and," the paper added, "he will not solve it, at the rate of his progress thus far, in a century." Because he had invented the phonograph, he now was being given the benefit of the public's abiding doubts; but his current claims were actually "childish and inferior... When will the public cease to believe a thing just because Edison says it is so?" Far from being the Wizard of Menlo Park, Edison was really the "Professor of Duplicity and Quadruplicity."[18]

Criticisms of Edison created two new images of the inventor: thief and businessman, which often were conflated. Reports of numerous lawsuits by rival inventors suggested that Edison was borrowing or stealing ideas; the multiplicity of his own corporations stood as evidence that he was in league with robber barons. Edison's association with Gould, Vanderbilt, Morgan, Villard, and other "vampires of the Stock Exchange" made it seem unlikely that he was interested purely in the pleasure of inventing.[19] In the face of such articles, Edison, for a few months at least, became uncharacteristically reticent about premature announcements of success. Once he had allowed reporters to publicize every step of his work, and was not above bribing them—sometimes with stock—for press coverage.[20] Even reporters who were not rewarded financially found themselves welcomed to the staff's midnight suppers and songfests; many hovered around Menlo Park. Now, as Edison struggled with the incandescent lamp, he appeared secretive.

Throughout the fall of 1879, as reporters listened for rumors of the Menlo Park experiments, Edison and his staff focused on carbonized cotton thread as a possible filament. Others had tried carbon, and Edison had rejected the material early in his experiments because it burned up so quickly. But Charles Batchelor and Francis Upton developed a way to process carbonized thread and shape it into spirals to make it more resistant. It was the first step toward success, and yet even this innovation was too fragile to be marketable. For two intense months, Batchelor and Upton experimented with more than 250 materials, including drawing paper, fishing line, cardboard, shavings from various kinds of wood, and cotton soaked in tar. Finally, in mid-November, a thin ribbon of carbonized cardboard, shaped into a horseshoe, glowed brightly for more than thirteen hours.

Menlo Park had its light, and in a notebook entry a few months later, Upton drew a caricature of the new bulb as a bespectacled man, smiling broadly, lifting his arms in triumph and apparently jumping for joy. "I shed the light of my shining countenance for $15,000 per share," Upton scrawled beneath. It was, perhaps, the most blatant use of the lightbulb as metaphor for an idea.

Although Edison still seemed reluctant to proclaim success publicly, he allowed Marshall Fox, a reporter for the *New York Herald*, to come to Menlo Park, making him promise not to write about what he saw until Edison gave permission to do so. With some backers offering $3 million and then $5 million if the lamps worked as a system, Edison was pressed to give a demonstration. He installed some lamps in his own home and that of his newly married assistant Francis Upton, whose wife bedecked the lamps with ribbons and lace, and he wired both houses to a generator. The result was nothing like the glaring arc lights: these lights were even and steady, and instead of the "ghastly hue" of arc lights, gave off a mild glow. There was no doubt that they were, Fox thought, undeniably beautiful.

Edison may have been naïve, or perhaps wily: Fox failed to honor the agreement and broke the news on Sunday, December 21, 1879, with an ecstatic headline: "Edison's Light. The great inventor's triumph in electric illumination. A scrap of paper. It makes light without gas or flame, cheaper than oil." As the article described it, Edison the conjurer had rallied: the light was magical, "incredible," a feat of alchemy. Edison had teased light "from a little piece of paper—a tiny strip of paper that breath would blow away." This ethereal thread emitted no gases, no odor, no smoke; "vitiating no air and free from all flickering" it was "a little globe of sunshine, a veritable Aladdin's lamp."[21]

The report in the *Herald* was tempered by other articles that emerged in the next days. On December 28, 1879, headlines on the front page of *The New York Times* expressed the simultaneous enthusiasm and doubt that characterized announcements in many newspapers and magazines: "EDISON'S ELECTRIC LIGHT, Conflicting Statements As To Its Utility." The *Times* reporter, more restrained than some of his colleagues, described Edison as a "short, thick-set man, with grimy hands," confident in his success, certain about the usefulness of his invention of the incandescent bulb, and dismissive of any criticism that electricians might level against him. "No electricians have been here yet," he told the reporter, "nor are there likely to be any here. Electricians are a very scarce article in this country, although there are many persons here who call themselves electricians." They stayed away, he said, because they did not want to admit that Edison had invented a "perfect" light. "Practical men, with experience, and what I call 'horse sense' are the best judges of this light," he added, "and they are the men whom I like to welcome to my laboratory." Henry Morton, president of the Stevens Institute of Technology, was certainly one of the experts, but when Edison learned of Morton's skepticism about this invention, he deigned to invite him to Menlo Park. Morton declined, but he did set out two challenges: proof that the carbon filament was as durable and that electrical lighting would be as inexpensive as Edison claimed.[22] Morton's expert opinion carried little weight: by December 28, a share in Edison's company, which had been selling for a few hundred dollars just weeks before, was up to $3,000.[23]

Edison was good copy for journalists, but besides creating an image of the inventor, they created an image of their readership: men and women expectant for novelty and disappointed, even indignant, when novelty was slow to come. This readership may have characterized reporters, who after all traded on news, but not necessarily the subscribers of *Harper's* or *Scribner's Monthly* or the *North American Review*. The desire for spectacle and amazement that had attracted eighteenth-century audiences appealed just as strongly to later generations. Yet if they enjoyed being dazzled, still, considering their reception of the telegraph, telephone, and phonograph, they were not clamoring for new technology. But beginning in the late 1870s, as the quest for an incandescent bulb accelerated, articles and reports about Edison—the public's primary means of knowing what he was up to—promoted an image of the public as ardently, anxiously impatient for change.

The media's hyperbolic reports of Edison's success, their relief that, finally, the incandescent bulb was a reality, seems out of proportion to the meaning of the invention for most of the public. As one letter writer to *The New York Times* noted reasonably, even if the incandescent lamp did work well, it would be "like having an elegant carriage without any horses... A man will not buy a pipe if he thinks he cannot buy tobacco to fill it."[24] What was the use of an incandescent lamp for most of the public, who had no access to electrical power, and for whom a different way of lighting city streets would mark no huge change in their lives or well-being? It is likely that this writer spoke for more of "the public" than did the effervescent reporters.

MARY ANN

The Menlo Park staff, of course, was working not for individuals without horses or tobacco but for its backers and the prominent businesses that would benefit from the lighting system. One of the staff's most crucial developments was a dynamo powerful enough to sustain the December demonstration: the "Edison Faradic" as it was called at first; then simply "our Dynamo." Its size varied as it was developed, but its configuration did not: in its most intimidating form, the dynamo had husky twin magnetic poles that stood over five feet high and weighed more than five hundred pounds each, with an armature that revolved in the cylindrical space between the poles. Although the current from the armature, as one technical journal described it, "could be used to excite the field-magnet," Edison had constructed another, separate, machine for that function.[25] The dynamo was massive, Amazonian, and the staff awarded it a suggestive, alluring, girlish nickname: Long-legged Mary Ann.[26] The Mary Ann dynamo—or "dynama," as Edison feminized the term[27]—was far stronger than the generator that Wallace had shown him in 1878, strong enough to light a building or a factory.

A few weeks after the investors' visit, Mary Ann powered a more public demonstration of Edison's lights, staged dramatically on New Year's Eve. As visitors pulled into the Menlo Park station, Edison's house and Upton's glowed "like a fairyland," a place of enchantment. But the spectacle quickly lost its luster when Edison revealed that neither his incandescent bulb nor his system was ready to leave the workshop. Once again, in the name of their readers, reporters expressed their disappointment. "Mr. Edison ought to know that the public will not be disposed to put its faith in him much longer if he does not take prompt measures for narrowing the gap between the fullness of promises and his lack of performance concerning the electric light," the *New York World* complained within days. "It is extensively believed that (in the language of lamps) he has turned the light alternately on and off the progress of his invention in a manner unpleasantly favorable to the manipulation of gas stocks."[28]

Two weeks later an "entirely disinterested gentleman," representing a city government that was considering investing in incandescent lighting, visited Menlo Park to assess Edison's progress. Prepared to witness an efficient and energetic laboratory and manufacturing center, he instead discovered workshops in disarray, workers engaged in testing telephones and phonographs, and a single glassblower whose output of bulbs was pitifully low. Even more astounding, he learned that the New Year's Eve display had consisted of a mere thirty-four incandescent lamps, supplemented by oil lamps "that helped to dazzle the spectators."[29] A month later, *Nature* agreed that Edison's premature announcements make it "harder than ever to trust him."[30]

But if such articles fueled disillusionment and discontent among readers, Edison, publicly at least, dismissed what he saw as uninformed gossip. Nevertheless, the pressure of work and, no doubt, the media took its toll on Edison's health. This time, instead of concocting elixirs for his workers, he put together a mixture of peppermint oil, chloroform, morphine, and alcohol to relieve his suffering from facial neuralgia, or nerve spasms. And instead of keeping the potion on a shelf, he traded on his reputation and sold it to a few promoters. "Edison's Polyform," advertisements claimed, created electricity when it came in contact with the skin, and could cure "neuralgia, toothache, nervous headache, or any nervous disorder."[31]

In the fall, a new dynamo, the mammoth offspring of Mary Ann—this one nicknamed "Jumbo," for P. T. Barnum's traveling elephant—was ready for use. And shortly afterward, the Edison Lamp Works, one of eight companies formed by Edison to supply and install his lighting system, began to manufacture lamps. On December 20, at five thirty in the evening, the moment when Charles Brush lit his arc lights along Broadway, a train pulled into the Menlo Park station bringing several New York City aldermen and assorted investors and dignitaries for a private demonstration of Edison's incandescent lighting system. Some three hundred lamps "shone steadily and without the least painful glare, and were beautiful to look upon," *The New York Times* reported in an article that ran alongside of the announcement of Brush's successful installation. Edison, who greeted his guests "with all the frankness of a pleased schoolboy," patiently explained the advantages of his lights over gas, apparently ignoring the arc light entirely. Since he knew that his visitors were worried about costs, he promised that electricity would cause no leakage and a newly invented meter would insure that customers paid only for what they used; and of course he promised safety, too: "It is very easy for a man to go to his hotel, blow out the gas, and wake up dead in the morning. There is no danger of a man blowing out the electric light." After touring the workshop, the machine shop, and Edison's library, the group returned to the laboratory, where they were treated to a sumptuous dinner of turkey, duck, chicken salad, and ham, and ample bottles of wine, all spread out for them "beneath the light of the electric lamp."[32] This public was pleased.

In April 1881, the city of New York, over the mayor's veto, granted Edison a franchise for laying wires in the Pearl Street neighborhood of Manhattan, where the offices of Drexel, Morgan, the *New York Herald*, and *The New York Times* were located, along with many other important businesses and some slum housing. Between the fall of 1881 and June 1882, Edison's company wired buildings and laid conductors in the streets, following gas lines. On June 29, they fired up boilers so that each of the six dynamos could be tested. The staff worked day and night, testing one dynamo after another, switching current to the feeders, testing the lamps connected to the current. By the end of August, they were convinced that all was running smoothly, when a police officer appeared with reports that horses driven past the corner of Nassau and Ann Streets were behaving strangely. The horses, shocked by an apparent leak, fueled the concerns of what Edison's team called "the worry hunters," who called for an improved system of underground conductors.[33]

Edison counted himself among the worriers. "I felt a great sense of responsibility," he said years later, "for unknown things might happen on turning a mighty power loose under the streets and in the buildings on lower New York. However, I kept my counsel."[34] Perhaps his worry caused him to keep the progress on Pearl Street quiet; only a dozen people gathered on the afternoon of September 4, when his first central station went into operation and four hundred lamps burned simultaneously for the eighty-five customers who agreed to have their buildings wired. Unlike Morse almost forty years earlier, Edison did not feel an irresistible impulse to invoke the divine. "I have accomplished all that I promised," he finally was able to say. A new light, apparently, had dawned.

DANCING LIGHTS

Three years later, the Stout-Meadowcroft Company advertised Edison electric lamps for home use. These were portable contraptions, since few homes were connected to a central station. In fact, despite aggressive marketing on the part of the Edison companies, just three villages contracted for their centers to be wired: Roselle, New Jersey; Sunbury, Pennsylvania; and Brockton, Massachusetts. Although Edison's agents had given estimates for wiring eighty more towns, only twelve had signed contracts, and even in these towns, electricity would reach the small number of residents living close to the generator. Still, if a homeowner could afford them, battery-powered electric lamps were available. The various parts were expensive: 8- to 16-candle bulbs sold for $1.50 each and needed to be replaced frequently, since the bulbs blackened and then burned out; a socket added an additional 25¢; a lamp stand with a shade cost $5; and a reflector, to augment the dim light, was $1. The most expensive part of the lamp was the battery to which it was hooked up: for $12, a battery would power an 8-candle lamp for an hour and a half. Even Edison's workers, who earned up to $800 a year, could hardly afford the investment. Only the very rich—J. P. Morgan or Cornelius Vanderbilt II, for example—bought generators for their homes. Although Morgan delighted in the electrification of his Manhattan mansion, Vanderbilt's experience was less delightful: in the picture gallery, the silk wall covering, woven with tinsel, burst into flames when crossed wires ignited. The conflagration was Mrs. Vanderbilt's first clue that a generator had been installed in the basement, and she insisted it be removed.[35] At a costume ball in 1883, however, Alice Vanderbilt arrived dressed as "Electric Light" in a costume that involved abundant diamonds, a concealed battery, and a torch.

Newspapers frequently reported electrical fires and accidental electrocutions, and magazines offered long lists of cautions for those who dared to install electricity. Prominent authorities such as Harvard physics professor John Trowbridge warned that stringing electrical wires above ground could lead to "disastrous conflagrations" if the wires came into contact with other wires or woodwork during a thunderstorm.[36] Because incandescent light had a different quality from gas, some people worried about becoming blind from reading by electricity. "A new disease, called photo-electric ophthalmia," reported the journal *Science* in 1889, "is described as due to the continual action of the electric light on the eyes. The patient is awakened in the night by severe pain around the eye, accompanied by an excessive secretion of tears. An oculist of Cronstadt is said to have had thirty patients thus affected under his care in the last ten years."[37] Women worried that electric light would produce freckles; in any case, they found that they looked more attractive in the mellow glow of gaslight.

Electric companies, including Edison's, were busy trying to convince potential users that all wires would be insulated and that safety fuses would cut off the current if overheating occurred, but still they could not counter the fear that nature would exact retribution for harnessing its power. Robert Hammond, in a book about the future of electricity for home use, reported a conversation with a man "who . . . told me very seriously that he quite expected these electricians would be put down by law, because if they consumed as much electricity of the air as they proposed to do, they would upset the balance of nature, and probably jeopardize the existence of vegetable and animal life upon the earth."[38]

Besides needing to quash public concerns about safety, Edison was competing for customers not only against other electrical companies but against the gas industry, which quickly offered new and improved burners, such as the "Incandescent Gas Light," developed by the Austrian inventor Carl Auer. The burner, or gas mantle, consisted of a fine cotton fabric impregnated with a solution of oxides and dried. As the mantle was used, the cotton itself burned off quickly, leaving a delicate network of oxides that burned longer, produced a brighter light, and gave off less heat and fumes than ordinary gas burners. For many consumers of gas, this improvement was significant enough to stall consumption of electrical power. Users were unaware that the mantles were produced in factories lit by electricity.[39]

Gaslight was both familiar and reliable. In 1884, *Scientific American* reported that the "novelty of the new light has worn out, and the extent to which it is being introduced as a substitute for gas is little noted except in the reports of the companies." Illumination of New York's Madison Square offered evidence of the kinds of problems that made the public wary. In a rainstorm, lighting was erratic, even though, in its own way, beautiful. "The lights danced up and down," one observer reported, "varying with the floods of rain apparently, occasionally sinking to a dull red, and then going out altogether and leaving the wind swept square in total darkness. Then the lights would flash out gloriously, flooding the spaces with their dazzling brilliancy and defying the elements that raved through the air. And so, up and down, the rays of electric lights rose and fell through the tempest, and they who were fortunate enough to see the show without being exposed to the wild storm will long remember the spectacle."[40]

That kind of spectacle, though, was not helping to market electricity; and since electrical wiring offered nothing but the possibility of incandescent lamps, consumers were not motivated to make the change from gas. In the Stout-Meadowcroft catalogue, only one new appliance was featured: the Iceberg electric fan, costing $7.50, and requiring a battery, at $13.50, which would run the fan for ten hours. At fairs and expositions, visitors saw models of electric sewing machines and stoves, but throughout the 1880s, these appliances were only a futuristic dream. Instead, at dozens of exhibitions held throughout America and Europe—from Louisville, Kentucky, to Vienna, from Caen, France, to New Orleans—visitors thrilled to dazzling displays of electric lights, usually installed by Edison. At London's Crystal Palace exhibition in 1882, Edison hung a chandelier, fifteen feet high, ten feet in diameter, embellished with ninety-nine incandescent lamps. The chandelier was designed as a bouquet, with flowers of brass and colored glass, a lamp bulb glowing at the center of each flower's petals. "The foliage is all of hammered brass, richly gilt... The corollas of the flowers containing the lamps and acting as their shades are... made of glass" tinted in pearl, white, ruby, clear olive, noted one admiring reporter.[41] The stunning chandelier portended the future of electric light: aesthetic, beautiful, derived from and even excelling nature. Nearly three hundred thousand people attended the International Electrical Exhibition in Philadelphia in 1884, where Edison's work was featured; and at the Paris Exposition of 1889, Edison's display, with ten thousand lights, illuminated an entire acre.

All of this luminous grandeur, meant to whet the public's desire for electricity, served mostly to perpetuate the public's love of spectacle. If electricity were to transform one's life drastically, as the fairs and expositions promised, how would that happen? Surely not simply by dazzling. In "Electricity As a Factor in Happiness," the author acknowledged the "anxious and hopeful attention" focused on electricity while the world, in 1881, waited "on intellectual tiptoe." What the world waited for, the author said, was, first, perfection of inventions already being heralded as life changing. Neither the telegraph nor the phonograph was reliable; and as for lights, "the big electric lights flicker and go out unexpectedly, and the little lights are not as bright as they should be, and all the lights are more or less disagreeable in color, and nobody will give you the least dependable hint about cost, and everybody tells a different story about the distance at which the force begins to tire and slacken." Even if these inventions were perfected, what, the writer asked, "will be the addition to human happiness? It is always necessary to ask that question, for, as a rule, the grand prizes of human intelligence, the additions to human knowledge of which we are so proud, have added little to the happiness of the millions who, and not the few rich, constitute man."[42] Would electricity bring peace, ensure equality, free people from oppression, banish terror? Or would wires strung across public squares leak electricity into the atmosphere, upsetting the balance of nature? Would electricity have dire consequences to the body or the mind, as yet unforeseen? Those were the questions that remained, for the world in the 1880s, unanswered.

CHAPTER 4

NERVE JUICE

George Beard at Yale, 1862.

Manuscripts and Archives, Yale University Library

We cannot create mechanical force, but we may help ourselves from the general store-house of nature.

Hermann von Helmholtz

In October 1874, Thomas Edison received a note from inventor Jarvis B. Edson, who recently had become one of Edison's partners in the Domestic Telegraph Company. Edson wrote to introduce a prominent young physician, George Beard, suggesting that he might be useful to Edison in promoting his new invention, the inductorium, an electrotherapy instrument featuring induction coils, which he was marketing as a reasonably priced alternative to other machines. Besides buying an inductorium for his own practice, Beard might well endorse it, Edson said, noting that "a certificate from him would be of great value—as he is an acknowledged authority on such matters."[1] Beard and Edison soon corresponded and discovered that they had much in common besides interest in electricity: boundless ambition, a talent for self-promotion, and audacious self-confidence. By the fall of 1875, Edison shared with Beard investigations he was conducting on an anomaly that both men believed could bring them lasting fame. Although they considered themselves kindred spirits, their partnership, nevertheless, was surprising.

George Beard was born on May 8, 1839, in Montville, Connecticut, the last of four children. His father, Spencer Beard, was a Congregational minister who raised his family, a friend of Beard later wrote, "in all the strictness and strait-laced orthodoxy of the times." Spencer Beard's Calvinism created an atmosphere so joyless that it tamped out young George's natural exuberance and eventually produced resentment of both the father and his theology.[2] It is likely that the somber atmosphere was exacerbated by the death of George's mother shortly after his third birthday. Family letters suggest that the young child was sent to live with relatives until, a year later, his father remarried. George's stepmother, Mary Fellows, herself the daughter of a Montville minister, did nothing to brighten the atmosphere at home. The children were raised to believe that only God could insure their success in life; their task was to worship and serve the deity. The two oldest boys, Spencer and Edwin, accepted this mission docilely, but it was more difficult for George, who resisted the strictures that his family forced upon him, yet at the same time wanted to please his parents and earn their praise. He was sent to Phillips Andover Academy, taught for two years after graduation, and then enrolled at Yale, with the family's hopes that he would follow his father's calling and become a minister, the career to which his two brothers were headed, but a career to which George felt no attraction.

Growing up in an essentially fundamentalist family, Beard experienced firsthand the transformation of psychological stresses into somatic symptoms. By his own account he was a depressed young man, and his depression manifested itself in assorted physical ailments: dyspepsia and other digestive problems, ringing in his ears, and recurring loss of vitality. He was convinced that he simply was not a good Christian; he did not believe deeply enough; he was not filled at every moment with gratitude toward God. "As I look back on my life," he wrote at nineteen, "I feel mortified & humbled at the little progress I have made in the Christian life." He read the Bible daily, made earnest resolutions to be a better person, and, besides his schoolwork, taught a weekly Sabbath class and went door to door selling subscriptions to a magazine called *Family Devotions*. Still he exclaimed in his diary, "O! that I might better improve the coming year that every day may be pregnant with good deeds;—that at my next birth day I may be better prepared for the battles and trials of life & for the solemn hour of death."[3]

One of those trials was the atmosphere of religious skepticism that he found at Yale; instead of offering intellectual liberation, the open and often contentious questioning of religious tenets distressed him. He felt "sometimes in doubt as to my own existence... I sigh oftentimes for the humble, trustful faith of my younger days, the zeal, the confiding earnestness."[4] As before, his doubt generated physical symptoms. He consulted a physician who prescribed some medicine, advised him to take a cold friction bath each morning, and told him that he "must avoid close mental application,"[5] advice that was difficult to follow for the competitive, ambitious undergraduate. Instead, Beard decided to stop eating meat and generally follow a temperate diet. This change helped, but his anguish about not living up to his parents' expectations still exacerbated his depression, severely enough that he sought treatment by a lay electrotherapist who practiced in New Haven. One roommate remembered the "magnetic battery" that Beard kept to administer electricity on his own.

Beard hid his anguish well. Except for a few close friends, most who knew him noted his "even and sunny disposition and . . . imperturbable good nature."[6] He was successful at Yale, and engaged in both studies and extracurricular activities. He exercised vigorously in the gymnasium and served as editor of the *Yale Literary Magazine*. He read poetry and novels in his spare time: Dickens, *Jane Eyre*, *Confessions of an English Opium Eater*, Hawthorne, and James Russell Lowell. Yet as his parents saw it, his success was caused by God's will and all could be dashed in a moment by the "trials and disappointments" God might inflict upon their son "to restore you to Himself."[7] Even when Beard was still a student at Andover, his father had warned him that success put him in peril; all good Christians, he said, must guard against "immoderate elation of too great self dependence or of an over estimate of our prowess, attainment or goodness."[8] When Beard thought about his future, his excitement was tempered always with anxiety. "I look forward to life with buoyancy and eagerness," he wrote in his diary after deciding to become a physician. "How long will it be before I shall be humbled & crushed? How long before I think more of bread than of fame?!"[9] If he thought of himself and his own desires, he was sure to be doomed: "George do you love this Savior?" his stepmother wrote to him. "Is he *your* refuge *your* portion *your joy your* strength? Is he near at hand not a God far off? Can you say truly *My Lord and my God*?"[10] He tried, but he could not answer those questions affirmatively, and often he felt overwhelmed by guilt and self-recrimination.

Knowing that he disappointed his family by not studying for the ministry, worried about the consequences of his decision, still he was intent on pursuing medicine, for which he had both passion and, he thought, genius. He knew his family disapproved: if there was healing to be done, it should be directed to the soul. After his brother Spencer became a minister in 1862, Beard commented sardonically in his diary: "Spencer is just licensed—ready now to launch faith on the world."[11] Both brothers, Beard thought, lacked the necessary energy and strength of presence that would make them effective ministers. Still, they, and not he, were fulfilling the family's dream. Launching faith on the world should have been his destiny, and, as his family saw it, he rejected that destiny at his peril.

Nevertheless, Beard persisted in following his dream: he was "born to be a physician."[12] After graduating from Yale in 1862, he enrolled at the Yale Medical School. After a year of studies there, however, and a semester at the College of Physicians and Surgeons in New York, he joined the navy, serving as acting assistant surgeon. During eighteen months of service, he spent nine consecutive months on a ship off the coast of Louisiana, "abundant opportunity," he said, "of testing Dr. Johnson's remark, that a 'ship is a prison, with the additional chance of being drowned.'"[13] Besides medical duties, Beard continued to write, contributing pieces to the *New York Tribune* and the *Times*, and two Boston papers. When he was ordered to serve on a flagship in South America, though, he happily resigned and returned to medical studies at the College of Physicians and Surgeons, from which he graduated in 1866.

Beard was engaged at the time, planning to marry as soon as he established a medical practice. Settling in New York City, however, made his goal difficult. Like many urban centers after the Civil War, New York did not lack medical practitioners offering a variety of conventional and alternative therapies, and Beard struggled to compete for patients and earn a steady income. So did a friend of his, Alphonso David Rockwell, who had studied medicine first as an apprentice to a Milan, Ohio, physician and later at Bellevue Hospital Medical College in

New York. Like Beard, Rockwell had served as a surgeon during the Civil War and then returned to Manhattan, where he opened an office. His education, Rockwell said, "had been very superficial and unsatisfactory... and in some branches [for example, obstetrics] I had absolutely no practice." His lack of skill coupled with his lack of confidence was hardly a prescription for success. He watched with envy as William Miller, a practitioner in an adjacent office, treated a steady stream of twenty-five to thirty patients a day, charging $1 for each visit. Miller called himself an "electrician" and focused his entire practice on applications of electrotherapy. Rockwell was fascinated, partly envious of Miller's success, but also because he wanted to understand more about Miller's therapeutics. "It must be remembered," he wrote later, "that at this date the whole subject was a veritable *terra incognita*, and to touch it, as one worthy friend remarked to me, was to imperil one's professional reputation."[14] For Beard, who loved a good fight, taking on the medical establishment seemed a decided attraction. "I am willing, I desire to have opposition if it comes legitimately," he admitted exuberantly when he was twenty-two. "In fact, I covet it."[15] In 1867, when Miller retired, Beard and Rockwell decided to take over his practice.

Electrotherapy was not quite terra incognita. Although electricity had long been part of the arsenal of mainstream medicine, lay practitioners, some of doubtful reputation, also administered electrical treatments. Physicians who practiced electrotherapy faced a special challenge in disassociating themselves from quackery and charlatanism. Yet even quacks achieved results, Beard saw, and patients wanted results, from whatever form of treatment. "The fact that such ignorant, reckless, unprincipled charlatans have been so successful oftentimes," he said, "is itself powerful evidence that *electricity* is an agent of very great efficacy in the treatment of disease."[16]

Electrotherapy reflected a new emphasis on the effect of the environment on the individual. Generations of physicians had prescribed change of climate to cure certain diseases, sending their patients to spas, to the seashore, or to mountains; generations of patients believed that certain climates bred certain diseases. But the thinking about environment that emerged in the mid-nineteenth century was different, emphasizing the importance of one's immediate surroundings, an environment that each individual could shape and control. A patient did not have to leave home to get well, but only had to create a home of cleanliness and serenity, a refuge where one could resist external demands on one's strength. "The natural, healthy condition of the mind is that of unruffled calmness," reported *Scientific American*. "If excitements occur, they should be exceptional, not the rule of life. As soon as they become a necessity there is a diseased state of mind and body, and the candle begins to burn at both ends."[17]

This shift of interest to one's immediate surroundings placed some responsibility for cure in the hands of the patient and not exclusively on drugs prescribed by the physician. In 1869, an article summing up "The Present Status of Medical Science" predicted that drug companies had little chance of surviving to the end of the twentieth century. "Before the expiration of that period, man will, perhaps, not have practically learned that diseases may be warded off by a clean, temperate life; but he will, at least, have learned that diseases, once acquired, cannot be cured by cathartics, emetics, or any of the other 'ics,' and, throwing himself upon nature, will give her the best chance to work he can, and thus secure the only possible chance he has for recovery."[18] The eminent physician Oliver Wendell Holmes certainly agreed: if the entire materia medica were thrown into the sea, he said, only the fish would suffer.[19] Skeptical of the effectiveness of drugs, patients were eager for other means to cure illness and maintain health. Invigoration by electrotherapy was one; restoration through rest was another.

The best-known proponent of therapeutic rest was Silas Weir Mitchell, a contemporary of Beard's, who had graduated from Philadelphia's Jefferson Medical College in 1850 and served as an army physician during the Civil War, treating soldiers suffering from exhaustion, nerve injuries, gunshot wounds, and amputations. When he resumed his full practice after the war, Mitchell noticed that many of his patients seemed to be suffering from a kind of battle fatigue, and he concluded, like Beard, that nerve weakness was the cause. The cure, according to Mitchell, consisted of enforced rest, regulated nourishment, and carefully monitored regimens of exercise. "The moral uses of enforced rest are readily estimated," he wrote. "From a restless life of irregular hours, and probably endless drugging, and from hurtful sympathy and over-zealous care, the patient passes to an atmosphere of quiet, to order and control, to the system and care of a thorough nurse, to an absence of drugs, and to simple diet."[20]

The rest cure, as it came to be known, was more complicated than Mitchell's summary implies. Although it effectively separated patients from the stresses of daily life, it also deprived them of autonomy and of the compassion that Mitchell deemed "hurtful." From the "over-zealous" ministrations of caretakers not under the physician's watch—presumably family members—patients gave themselves over to "a thorough nurse" who was charged with enforcing the physician's orders; although patients were freed from soporific drugs or alcohol-based tonics, they were forced to eat as many as eight meals a day, meals notable for starches and fats, whether or not they had any appetite. Their job was to gain weight. And although "an atmosphere of quiet" and "order" might sound restorative, that atmosphere precluded any activity at all. Books, pens, and paper were forbidden. Visits with family or friends were forbidden. A patient's exertion of will was forbidden—until the physician deemed the patient cured.

Mitchell prescribed his rest cure for men and women, but the plight of some of his famous female patients—Charlotte Perkins Gilman, Jane Addams, and William Dean Howells's daughter Winifred, for example—has given Mitchell a posthumous reputation of oppressing women. Certainly he shared his contemporaries' assumption that a woman's highest achievement should be as wife and mother, but he also believed women should have an opportunity for education, and he recognized that the central problem in some of his patients' lives was an overbearing or insensitive husband. Mitchell himself, suffering from many of the symptoms of nerve weakness that plagued his patients, claimed to have undertaken the same kind of rest cure that he prescribed for others.[21] The rest cure resulted less from Mitchell's sexism than from his beliefs about how physicians must exert their authority, authority that patients did not automatically confer upon medical practitioners.

While Weir Mitchell's fame is indelibly connected to his rest cure, he also enhanced the profession's status by contributing to a new genre of writing: self-help medical guides. His *Wear and Tear* (1871) and *Fat and Blood* (1877) were hugely popular books offering advice on healthful living. Physicians who wrote such books had no intention, of course, of driving away clients, but instead of elevating the status of the physician by underscoring professional expertise and by introducing readers to a plethora of illnesses for which they could seek professional treatment.

Beard's *Our Home Physician*, published in 1869, spoke directly to this goal and also, as his choice of pronoun indicates, to present the physician as the patient's ally and supporter. One purpose of his book, Beard wrote in the Preface, "is not so much to enable its readers to dispense with a *physician*, as to teach them how to dispense with *disease*." If patients did not suffer disease, any reader might ask, why would they need a physician? But Beard does not bother with this question, leaving perceptive readers to surmise that their power to dispense with disease might prove limited. In fact, Beard reveals another purpose for his book: to disseminate information about "the recent inventions, discoveries, and improvements by means of which physicians are now enabled to study and to treat disease so much more satisfactorily and successfully than in former times." These new "instruments and appliances . . . cannot, of course, be used by my readers," Beard cautions, but only by trained physicians. Yet one can see in his list of these innovations that only a few are beyond the capability of patients. "Instead of bleeding and calomel, tartar-emetic and low diet," Beard writes, "we now give *tonics and stimulants—iron and quinine, strychnine and arsenic, cod-liver oil and whiskey, air and sunlight, passive movements, general electrization, abundance of sleep, and a large and palatable variety of nourishing food*."[22] If patients needed a prescription for some tonics and stimulants, still they had access, as they always had, to sunlight and whiskey. But the new treatment embedded in this list could be found only in a doctor's office: electricity. Just as Mitchell positioned himself as the authority over the isolated patient undergoing a rest cure, Beard conferred upon himself the authority of the expert in manipulating certain "instruments and appliances": batteries, wires, wands, and brushes. His cure for nerve weakness was electricity.

THE EARTHLY IDEAL

More than twenty years before Beard embarked on electrotherapy, Hermann von Helmholtz had formulated what he called the law of conservation of force. From experiments in physics and chemistry, Helmholtz concluded "that nature as a whole possesses a store of force which cannot in any way be either increased or diminished; and that, therefore, the quantity of force in nature is just as eternal and unalterable as the quantity of matter." As physicians and their patients interpreted this statement, it implied that the human body could not generate limitless amounts of energy. Either the body needed to rest and conserve those energy stores it had, or the body needed to borrow energy—to nourish itself—from external sources. "We cannot create mechanical force," Helmholtz stated, "but we may help ourselves from the general storehouse of nature."[23]

Electrotherapy, Beard claimed, borrowed from nature's larder to nourish a depleted nervous system. "All these forces with which we are so familiar—light, heat, electricity, magnetism, motion, the vital force of plants and the nervous force of man—are simply *modes of motion*," he wrote in *Our Home Physician*.[24] Each person, he believed, contained a certain amount of energizing material; when demands on the body exceeded this amount, nerve weakness resulted. It was a straightforward matter of supply and demand. Beard's acclaim for this treatment, however, resulted from more than his understanding of human physiology: electrotherapy, he thought, was the key to professional success.

Beard's use of electrotherapy was not an innovation but a reformulation and expansion of prevalent medical practice. Even Mitchell administered electricity to make sure that his patients' muscles did not atrophy as they waited for release from the rest cure. Since the eighteenth century, physicians had used

Illustration from Beard and Rockwell: A Practical Treatise on the Medical and Surgical Uses of Electricity. *Physicians used galvanism to treat a large menu of common ailments.*
The Bakken Library, Minneapolis

electric shocks to treat paralysis, spasms, and convulsions, and to effect resuscitation in cases of drowning or suffocation; they claimed electricity's usefulness to break up blockages in the digestive, circulatory, or reproductive system. When twenty-six-year-old Henry James was traveling in Florence, suffering from acute constipation, his brother William recommended that he find an electrotherapist. Electricity—which Henry had tried at home the year before—might cure the condition, "applied not in the piddling way you recollect last winter but by a strong *galvanic* current from the spine to the abdominal muscles, or if the rectum be paralysed one pole put inside the rectum." Although William knew of no Italian electrotherapist, he was aware of several with excellent reputations in Europe: the Viennese physician Moriz Benedikt, Ernest Ominus in Paris, Julius Althus in London.[25]

In England, Guy's Hospital had established an "electrifying room" in 1836, where patients received treatment for chorea or other spastic disorders, paralysis, seizures, or hysteria. The goal of this electrotherapy, however, was to control the body's abnormal behavior and not, as Beard proposed, to nourish the weakened body. Although in an "electrifying room" some patients were offered treatment in the form of a mild "electric bath," many more patients were forced to submit to shocks from a Leyden jar or voltaic battery, shocks strong enough to jolt a paralyzed limb and painful enough to make any hysterical malingerers change their behavior.[26]

Electrotherapy, as Beard defined it, should not be punitive; the delicate process of administering electricity, therefore, made electrotherapy "the most exacting and laborious of all the special departments [of medicine], for in a certain sense it trenches on and necessitates a knowledge of all other departments." The physician must be an expert diagnostician, able to choose the correct electric current (faradic or alternating, intermittent current; galvanic or direct, continuous current), decide on the most effective method of application, the appropriate strength of the current, the number of treatments, and the type of apparatus. Clinical experience, Beard asserted, was the best teacher for physicians entering this complex area of medicine; but self-experimentation was important, too, in giving the physician an accurate sense of the patient's experience.

That experience could be intimidating at first, and even frightening. In galvanic or faradic treatments, the types most often prescribed, the patient's skin was moistened and damp fabric placed between the electrode and the patient's body to ensure conductivity of electricity. The physician fastened electrodes at the site to be treated—head, base of spine, uterus—and connected them to a battery by wires. As the galvanic treatment began, the patient would feel an electrical shock, muscles would feel stimulated, and as the treatment continued, provided it was monitored correctly, the patient would feel nothing as the current flowed through the body. Patients sometimes complained when the initial shock subsided, believing as they did that feeling a buzz or tingle meant that the electricity was working. Sometimes, however, they simply fell asleep during the treatment. Often applied for its sedative effects, galvanism proved a treatment of choice for nervous illness and insomnia. For the pain of neuralgia, galvanism, according to some patients, bordered "on the miraculous."[27]

While physicians believed that galvanic treatments caused chemical changes in tissues, faradic treatments, on the other hand, were thought to have a mechanical effect on muscles, which was best for treating cases of paralysis and an assortment of gynecological complaints, such as painful menstruation. Faradization was more shocking to the body, involving a rapid series of pulses, each feeling like a sharp electrical shock and producing a buzzing, shivering sensation. The intensity of faradization depended on the strength of the current and the frequency of the pulses.[28] One form of faradization favored by Beard was the use of the electric "hand" by which the physician administered electricity that coursed through his body to his hand. While the patient's feet rested on a piece of copper attached to a negative electrode, the physician held the positive electrode wrapped in a sponge, moving his hand over the patient's upper body and squeezing the sponge to vary the current. Male patients partially undressed for this procedure; women were shrouded in a sheet.[29]

Electric baths proved appealing for some patients. Unlike eighteenth-century "baths," which immersed patients in an atmosphere of electricity, these baths immersed patients in electrified water. Hospitals or clinics at spas constructed special cabinets for this treatment, in which electricity could be administered along with medicated vapors or steam. In the physician's office, baths required an insulated tub for full-body immersion or a wooden basin for treatment of a hand or foot. Electric baths could apply galvanic, faradic, or a combination of galvano-faradic current, thereby making them appropriate for a wide range of ailments. The patient sat in the bath while a weak current, too weak to cause harm and often too weak to be felt at all, passed through and electrified the water. After an electric bath, patients testified to increased appetite, decreased digestive problems, and a noticeable feeling of tranquility. The baths were so popular that patients often requested them instead of other electrical treatments. One physician reported that when he demonstrated faradization to a married couple so that the wife could supplement office visits by administering treatment to her husband at home, she protested that she, too, wanted electrotherapy; she "had a hankering after baths & Swedish massages," the physician noted, "& when I left them it was uncertain what her husband would decide to do under the strong influence which she evidently exercises over him."[30] Because in all forms of treatment applications of electricity could vary subtly, the physician had great influence in shaping the patient's experience and great latitude about deciding what ailments would respond to a series of treatments. Rarely could relief be effected from just one session.

Although in 1867 Beard had little experience as a clinician, still he was certain that electrotherapy would prove beneficial to patients and, furthermore, would bring to therapeutics in general "a wider liberality and a broader spirit of inductive investigation. The paths of all future explorers in the scientific treatment of disease will have been made easier and safer by the toils and the triumphs of this one department."[31] The possibility of triumph was never far from Beard's imagination. Reforming "medical & Hygienic knowledge," he wrote in his diary when he was still at Yale, "[t]his is my earthly ideal."[32]

A year after he set up his practice, Beard published advice to other young physicians, "The Practice of Medicine in a Pecuniary Point of View," outlining the four steps that, even by then, were proving successful for him and Rockwell. The physician, he said, should focus on practical results rather than scientific theory; patients should "feel their dependence" on the physician and never presume that their physician was dependent on them. To foster that dependence, physicians must be "frank, out-spoken, clear, and above all *positive*," Beard wrote. "Better let the diagnosis be positive even though it be wrong. Doubt is the practitioner's worst enemy." As far as fees, he advised that they be high: "Mean prices are apt to bring in mean patients; and the presence of mean patients deters and repels many who would pay large fees." And in closing he advised that physicians find "some department that is congenial to your tastes, and make yourself a necessity to the profession in that department" through research, publishing in professional journals, and "by personal influence."[33]

His first step in realizing that goal was in his own treatment rooms. In his first years as a physician, as he worked on building his clientele, Beard noticed a marked similarity in many patients' symptoms. In November 1868, for example, Mr. E.H., a forty-three-year-old executive in a fire insurance company, walked into Beard's office complaining that he had just experienced an attack of "cerebral congestion," about which he was extremely worried. But this attack was only the most recent of many other symptoms that Mr. E.H. had long been experiencing: dyspepsia, constipation, severe headaches, insomnia, and inexplicable fatigue. He would be exhausted after a walk; in fact, he said, he was frequently exhausted by his work, which involved "calculation, writing, [and] consecutive thought." Another man, twenty-eight and a physician, suffered nervousness and repeated headaches. "To use his own expression," Beard noted, "he had been 'living on a lower plane than was normal.'" Nineteen-year-old Mr. McC. could not sleep, had night sweats, and "aberration of memory." He was so depressed that he often thought of suicide. Mr. A, aged sixty, became exhausted after reading for more than ten or fifteen minutes; nor could he walk more than a dozen blocks without collapsing. What these men had in common, besides their symptoms, was lack of any organic illness. Except for loss of weight, they appeared healthy: no pallor, no abnormal pulse, no heart murmur. Yet their symptoms were severe and disturbing, and Beard, who had suffered from many of them himself, took them seriously.[34]

The symptoms, Beard believed—symptoms familiar to us as indicating depression and anxiety—constituted a special pathology caused by a weakened nervous system. These men—and sometimes, but not as frequently, women—were not hysterical; they were not, as some other physicians thought, malingerers. They were "brain workers" challenged by the stresses of their lives: working long hours, worried about their future, striving to make more money, pressured to push themselves beyond their body's capacity. Beard did not agree with prevalent medical assumptions that their complaints resulted from heredity, weak moral character, or the underdevelopment of some mental faculties; instead he believed that nerve weakness, which most often afflicted sensitive, intelligent individuals, was caused by overstimulation in a fast-paced, competitive environment.

Between 1867, when he and Rockwell became partners, and 1871, when they published jointly *A Practical Treatise on the Medical and Surgical Uses of Electricity*—a book that stayed in print for the rest of the century—Beard engaged in a campaign to publicize and legitimize electrotherapy and to diagnose one specific ailment for which it was most effective. In effect, he invented a new disease: neurasthenia.

THE PHILOSOPHICAL ANATOMIST

The term, long in the medical vocabulary, recently had appeared in an article by Edward Holmes Van Deusen, the medical superintendent at the Michigan Asylum for the Insane: "Observations on a Form of Nervous Prostration (Neurasthenia) Culminating in Insanity." Neurasthenia, or nerve weakness, seemed to Van Deusen more appropriate to his cases than "nervous prostration." The patients he treated at the asylum included depressed, isolated young farm wives, but also businessmen and professionals who complained of assorted symptoms, such as insomnia, night sweats, intestinal disorders, and, some said, bad breath. His patients' first and most striking symptom, Van Deusen reported, was "distrust." Those who were religious suddenly lost faith in "God's promises"; those who strived for financial gain were tortured by the prospect of ruin; those whose marriages were "particularly close and tender" were beset by jealousy. Fear of losing whatever one most deeply valued, anguish over a loss of control: these feelings characterized neurasthenia; and, Van Deusen asserted, neurasthenia could be caused by stress. "The hot-house educational system of the present day, and the rash, restless, speculative character of many of our business enterprises, as well as professional engagements," he wrote, "are . . . strongly predisposing in their influence to debilitating forms of nervous disorder."[35]

As Beard redefined the illness, neurasthenia encompassed a wider range of symptoms that applied not only to patients in an asylum but to practically everyone who might walk into his office. "The *diagnosis* of simple and pure neurasthenia," he claimed, "is arrived at mainly by *exclusion*." Frequently confused with anemia, neurasthenia was not organic, although it might be incited by organic diseases; the most common symptoms were depression and fatigue. While some patients might be predisposed to neurasthenia because of heredity, most cases could be traced to "special exciting causes," such as "the pressure of bereavement, business and family cares, parturition and abortion, sexual excesses, the abuse of stimulants and narcotics, sudden retirement from business, and civilized starvation, such as is sometimes observed even among the wealthy orders of society. *The disease is most frequently found in the United States, among the brain-working classes of our large cities.*"[36]

This last sentence, which Beard pointedly italicized, identified the patients he most wanted to treat: professionals, executives, office workers. Men and women who engaged in "intellectual exertion" that they found as exhausting as the most demanding physical labor. Men and women who could afford to visit their physician several times a week for months of therapy. The more intelligent person, Beard asserted, possessed a more sensitive nervous system, more prone to irritation and disturbance than the nervous system of someone less intellectual. Evolution, as Beard understood it, proposed a hierarchy from organisms with "lowly evolved (non-nervous) organizations" to complex, "highly evolved" organizations, and even within those complex organizations, some individuals were more sensitive than others; "reflex actions of all kinds take place in them more rapidly, in a far more complex way, and under slighter irritation; the echoes of nerve disturbance resound through every organ of a nervous man." In those individuals who are physically "strong and wiry," the body's resistance to the nervous action stops irritation before it spreads through the body, thereby causing localized organic disease; but in extremely sensitive individuals, irritation speeds through the nerves to cause generalized, systemic nervousness: "the active 'sympathetic,'" Beard claimed, "informs the whole system." In the evolutionary hierarchy, nervousness helps the system defy death. Irritation of the nerves cannot accumulate harmfully in any one organ or site, but dissipates as it scatters through the body. "[T]hus our very weakness becomes our strength and our salvation," Beard announced.[37]

Besides identifying the population susceptible to neurasthenia, Beard publicized the idea that "mental and moral diseases" were not limited only to the inmates of asylums but were common throughout America and Europe. Indeed, as he wrote in "Who of Us Are Insane?," "persons afflicted with the incipient and milder phases of what we call insanity are all about us, on every hand, and mingle with success in the various relations of life . . . [I]ts manifestations are most frequent and most severe in civilized communities, and among the intellectual or ruling classes." These manifestations—melancholy and nervousness, among others—were a result of overwork and overstimulation; just as dyspepsia was an illness suffered by those wealthy enough to overeat, insanity was an illness suffered by those important enough to be beset by other people's demands, motivated enough to strive for excellence, or, simply, so brilliant that insanity could hardly be distinguished from genius. While arguing for tolerance of a range of eccentric behavior, Beard also suggested that his readers might well examine their own behavior and ask whether it fell within the category of nervous illness, illness that could be treated by physicians who understood the consequences of nerve weakness and who respected patients who demonstrated "intellectuality and refinement."[38] The question Beard implied in his article was "Who among us are neurasthenic?"

CROWDS OF NEUROTICS

Many, he believed, and those who suffered required a long regimen of treatment. As Freud would realize when he began his own career, Beard saw that treating people with organic nervous disease meant a smaller practice than treating the "crowds of neurotics, whose number," Freud noted, "seemed further multiplied by the way in which they hurried, with their troubles unsolved, from one physician to another."[39] Rather than offer a "cure" by writing a prescription for such popular nerve tonics as quinine, iron, or phosphorus and not seeing the patient again, electrotherapists prescribed repeated visits, carefully monitored to determine whether the patient needed electricity's sedative or tonic effects.

Because the type of electrotherapy was fine-tuned for each patient, the physician needed to be sensitive and alert to how the patient presented symptoms. Beard, despite being partly deaf and despite his tendency to be bombastic in arguments, was considered to be a good and sympathetic listener; both he and Rockwell made a special effort to treat each patient as a distinct individual. "The distressed patient appeals to you to confirm his idea that never was there another case like unto or as severe as his own," Rockwell noted, "and while you reassuringly tell him, 'Many and worse,' yet if your observation is keen you will soon find that each case is a law unto itself, and that there is no stereotyped method of treatment."[40]

When Mr. A, for example, a retired businessman whom we would now call a workaholic, came to see George Beard in 1868, he was suffering mightily from persistent restlessness, insomnia, and fatigue. The cause of his symptoms was not overwork but a feeling of loss—of control, of power—that resulted from withdrawal from the business world. Even a short walk would cause him to collapse. He had tried every tonic his physicians had offered him, but nothing helped. Like other sufferers of such symptoms, Mr. A. had come to blame himself for failing to cure his ailments through willpower. He was hopeful and gratified when Beard presented both a different explanation for his patient's symptoms and a different cure. After many months of faradic and galvanic treatments, Mr. A felt stronger, more energized, and at the same time more relaxed; finally, the former captain of industry could enjoy his retirement.

Colleagues often referred patients whose symptoms they found intractable. Beard visited one woman, for example, unable to walk, unable even to leave her bed. She was clearly exhausted from nursing her five children through a serious illness, but slept only when she took "large doses of McMunn's elixir of opium." Beard brought batteries with him, applying "an exceedingly mild and fine faradic current over the head down the spine" for about five minutes. "The patient described the sensation as most delightful," Beard reported, "and expressed disappointment because the treatment was so brief." The next night she slept and within two weeks substantially reduced her reliance on McMunn's elixir. Treatment continued, of course; Beard generally gave treatments for a few months, then stopped to assess improvement without further electrotherapy. Some patients were sufficiently strengthened after a dozen sessions; others might need several hundred.[41]

Although Beard's patients, by his own account, were both grateful and cooperative, he still faced resistance when he submitted articles to professional journals. His first paper on neurasthenia and its treatment by electricity was rejected three times before the *Boston Medical and Surgical Journal* published it in 1869. The medical profession's response to electrotherapy resulted from more than concern that the treatment had been tainted by quacks; physicians, like the rest of the population, struggled with unanswered questions about nervous electricity, the relationship of mind to body, and the connection of mind and soul. If humans were energized by electrical force, if this force could be depleted and restored by artificial electricity, did that notion reduce the body to matter alone? How could one believe in nervous electricity and still believe that the soul, itself a possible manifestation of animal electricity, was the essential animating force? Electrotherapy as a treatment for nervous illness revived questions about vitalism and materialism that the nineteenth century had inherited from the past.

Beard did not avoid these questions as he defended his diagnoses and treatments. Two weeks before "Neurasthenia" appeared in 1869, he published in *The Medical Record* a review of *The Human Intellect* by Noah Porter, a former minister, and at the time Clark Professor of Moral Philosophy and Metaphysics at Yale; in 1871, Porter became president of the college.[42] Attempting to differentiate the mental activity of perception from that of imagination, Porter claimed that the soul is not dependent on matter but is a distinct entity imposed upon the body, which causes creativity, emotions—in short, personality. The soul, Porter maintained, "uses and depends upon the brain as its organ of communication with the material world," but is not generated by the brain or any other part of the body. Intelligence, sensitivity, genius: all of these are products of the soul. Opposing this view, Beard noted, were materialists, the outspoken Herbert Spencer foremost among them, who held that what others considered to be the "soul" could be reduced to the workings of the brain and nervous system. Chemical and physical processes caused an entity to be alive, creative, or brilliant; once those processes ceased, the individual would be dead, with nothing remaining but dead matter.

Beard wanted to distinguish himself from both views, and in doing so he formulated a theory that appears to reconcile the religious teachings of his youth with scientific evidence. Friends and colleagues generally agreed that Beard was an agnostic, finally rebelling against the religion imposed upon him by his family and the superstitions supported by their religious convictions. "Inwardly," Rockwell said, "he resented this long thralldom, and the pendulum which had swung so long in one direction made an equal arc in the other."[43] But Rockwell also acknowledged that Beard "worried about his soul a good deal, and scored himself for loving the vanities of life."[44] Beard genuinely may have searched for ways to affirm his religious legacy, or he may have been able to stop the pendulum to serve his purposes. In this review of Porter's book, his purpose was to embrace and not alienate those who would read his contributions about neurasthenia and electrotherapy, which were to appear frequently in *The Medical Record*. To explain the relationship among mind, soul, and body, Beard claimed, one needed to be a "philosophical anatomist." In every individual, "there is an immortal soul, which may be identical with the vital force, or may be correlated to it and to all the other forces of the body." The soul, moreover, "acting on the brain, produces intellect, just as when acting upon the digestive apparatus it produces digestion: in other words . . . the intellectual nature, in the *whole range* of its capacity to know, to feel, and to will, is the *function* of the brain, just as truly as digestion is the function of the apparatus of digestion." This neat explanation, Beard claimed, was borne out by research in comparative anatomy and, happily, was consistent with "Revelation, [and] with the intuition of mankind." His theory, he said, was "a compromise between the materialists and the spiritualists, and will, we think, in time be substantially accepted, at least by those who do not ignore religion."

If that theory was accepted, then Beard supposed his other ideas would be accepted, too: that neurasthenia was depletion resulting from the stresses of civilization; that mental diseases could be accounted for by organic changes in the brain; that intellectual labor produced the same kinds of chemical products in the body as did physical labor; that the "known and familiar forces—light, heat, electricity, magnetism and motion—are correlated to each other" and cannot be annihilated. Each individual soul, and God himself, is correlated to these enduring forces; and therefore the soul is immortal and God eternal. If

Porter had known science as well as Beard did, he might have made a different argument, Beard concluded generously, urging his own profession to fulfill "the great duty of diffusing its knowledge through society." In 1869, diffusing knowledge was one of Beard's primary tasks, which he hoped would yield the fame, respect, and, intermittently, fortune for which he had yearned. Edison's offer of collaboration seemed to open another path to those goals.

CHAPTER 5

SPARKS

The Wizard of Menlo Park. This front-page illustration in the New York Daily Graphic, *July 9, 1879, caricatures Edison searching for platinum to use as a burner in his incandescent lamp. His robe is covered with drawings of his inventions.*

U.S. Department of the Interior, National Park Service, Edison National Historic Site

An abstract idea of a natural law, I maintain, may be *invented.*

Thomas Edison

The project on which Thomas Edison and George Beard worked together beginning in the fall of 1875 portended, they believed, something greater than technological or medical innovation, something as great, in fact, as the achievements of Faraday or Ørsted. Edison claimed to have discovered a new form of energy that behaved in ways startlingly different from known forces. As Beard helped him test the effects of this energy on living organisms—expertise that none of Edison's muckers had—he, too, became convinced that Edison's discovery was monumental: "it compels," he said, "a reinvestigation of the laws and phenomena of the different forms of electricity," and therefore of the science that justified electrotherapy.[1] Both men hoped that the new force would have a personal impact as well: fame, and even glory.

As Edison worked on the quadruplex and acoustic telegraphs, he noticed something strange: sparks that emanated from a vibrator magnet in the absence of any electric current. He had seen such sparks in previous experiments with stock printers and even his electric pen, and he had assumed that they were caused by the generation of electricity by a magnet; but these new sparks seemed unusually strong and bright, and he decided to investigate them further. He connected a wire to the end of the vibrating magnet and drew a spark by touching a piece of iron to it. He connected a gas pipe and drew sparks from gas pipes elsewhere in the room, even though they were grounded. He found that he could draw sparks from a metal object placed in the vicinity of the vibrating magnet, even though no wire connected the object and the magnet. As any electrician then knew, electrical current needed to flow through a wire; these sparks seemed to be produced by electricity flowing through air. And there were other puzzles: oddly, the sparks did not register a charge on any galvanometer or electroscope, showed no polarity, and could not be felt.

The sparks presented a problem, and, for the twenty-eight-year-old Edison, an opportunity. Because he contributed frequently to the professional publication *Telegrapher* and served as science editor for the *Operator*, Edison had become known beyond the community of electricians. Just that summer, his name had appeared in newspaper reports about Western Union's new developments in quadruplex telegraphy; he had demonstrated some of his devices for the National Academy of Sciences in Philadelphia; and because of the prolific number of projects in which he was engaged—he had filed for nearly a hundred patents by the fall of 1874—reporters listened for news from his laboratory. Edison justifiably could be proud of his growing reputation, and yet he wanted to make a name for himself as more than a commercial inventor: he wanted to join the pantheon of great men of science, explicators of nature's hidden workings—a Galvani, a Volta, a Faraday. Being elected to the Society of Telegraph Engineers, a prestigious British association with a select membership, validated his achievements, but clearly he wanted his contributions to be seen as groundbreaking, not merely ingenious. He was planning to write a book about his experiments in telegraphy, a book not simply to sum up the state of current knowledge about the subject but to present some findings that he believed were unique to his own experiments.[2]

Beard, thirty-six, was at a point in his career where he hoped to gain an audience for his ideas that went beyond the medical community. In the past five years, his writings on neurasthenia and electrotherapy had gained him esteem, and sometimes notoriety, in his profession. But even before he met Edison, he had decided to devote himself less to his medical practice and more to writing and research. Like Edison, he, too, hoped to enhance his reputation. Beard coveted recognition, and Edison's sparks excited him.

The time was right, Beard maintained, for a new and significant scientific discovery. "It is forty-five years since any force has been introduced into science; the discovery of induced electricity by Faraday," he noted. The discovery of galvanism had come eighty-nine years earlier; the voltaic pile, seventy-six. Franklin had flown his kite 123 years ago, and it was 275 years since William Gilbert published *Tractatus de Magnete*, the treatise, Beard said, "that marks the birth of the science of electricity."[3] Clearly Beard believed that the new discovery could put him and Edison in the company of immortals. Beard suggested a name for the mysterious radiance: "apolia," he said, since the force apparently had no polarity. Somehow, however, that name was dropped in favor of "etheric," which Edison later attributed to Beard and Beard to Edison. "Etheric," of course, was steeped in history, suggesting the subtle fluid of ancient philosophy, mesmeric emanations, vitalism, and the Romantics' quest for a unifying natural force. It was an enticing and audacious choice.

Edison, unlike Morse, never cried out "What hath God wrought!" as a phrase to test his telephone or phonograph. He presented himself as a practical man who invented practical devices, impatient with theories, scientific or philosophical, unless they could be translated into tangible results. Yet there was another side of Edison that emerged in his scattered and sometimes contradictory remarks about spiritual and metaphysical matters, and in some investigations that did not result in marketable items, which reveals a tension between empirical knowledge and faith that he shared with his contemporaries. He was an admirer of the transcendentalist Emerson and the Swedish mystic Emanuel Swedenborg, whose works Edison had on his library shelves. Both writers celebrated the existence of a divine spirit that permeated all of nature, and Edison felt an affinity to their ideas.

Edison did not believe that investigating enigmatic phenomena would lead to reducing the inexplicable, imponderable, and incorporeal to physical or chemical reactions. He thought it possible to sustain beliefs in the inexplicable—immortality of the soul, for example—through those investigations. These beliefs were important to hold, Edison suggested, because they fulfilled a human desire for transcendence. "I am working on the theory that our personality exists after what we call life leaves our present material bodies," Edison once admitted. "If our personality dies, what's the use of a hereafter? What would it amount to? It wouldn't mean anything to us as individuals. If there is a hereafter which is to do us any good, we want our personality to survive, don't we?"[4] The immortality of the soul, the existence of a spiritual realm where personality persisted: these ideas suggest a divine, creative, ultimately unknowable spirit that imbued nature with meaning.

Edison's proposals for investigating nature's secrets were sometimes whimsical and even bizarre. One evening at dinner with George Lathrop, who was interviewing him for an article in *Harper's*, Edison, speaking "as if out of a deep reverie," described an elaborate project to inquire into nature: suppose, he said, an individual could control each atom of his being. "For instance... then I could say to one particular atom in me—call it atom No. 4320—'Go and be part of a rose for a while.' All the atoms could be sent off to become parts of different minerals, plants, and other substances. Then, if by just pressing a little push button they could be called together again, they would bring back their experiences while they were parts of those different substances, and I should have the benefit of the knowledge."[5]

In June 1878, *The New York Times* published a letter from Menlo Park, signed only with initials but apparently written by one of Edison's staff, suggesting that the telephone technology on which Edison was then working was part of a larger project that had little to do with personal communication and more to do with penetrating "one of the innermost shrines of nature's temple." Edison, the writer claimed, was involved in a project of "Kosmophonics," inventing instruments that would enable one to hear "the sounds of molecular vibrations" throughout the natural world. "For the graphical registry and subsequent study of the voices of the animal creation we shall have the zoophone... For the registry of the sounds of the 'vegetable kingdom' we shall have the phytophone, for the sounds of inorganic nature the physiophone. All these names which now appear in print for the first time will doubtless be as familiar to the rising generation as telegraph and phonograph are to our contemporaries."[6] In the context of Edison's other reveries, it is difficult to know if the letter writer meant to be facetious.

Edison rejected the idea of "discovery," which he defined as "more or less in the nature of an accident" and surely not as significant as invention. Some of his contemporaries who were praised as inventors were, as he saw it, merely discoverers. George Goodyear, for example, accidentally found a way to harden rubber while he was at work on a different project; Alexander Graham Bell, experimenting with sending sound waves over telegraph wire, "discovered that articulate speech could be sent over the wire—and there was the telephone." But neither of these men, according to Edison, set out to solve a particular problem through persistent trials, reformulations of theory, and "patient labor." "In a discovery there must be an element of the accidental, and an important one, too; while an invention is purely deductive. An abstract idea of a natural law, I maintain, may be *invented*." By his way of thinking, Isaac Newton had invented, not discovered, the theory of gravitation, a problem on which he had been working for years.[7] Surely Newton was more important to the history of ideas than Goodyear. Despite the accidental element in his latest discovery, Edison hoped to invent an explanatory theory that itself would be revolutionary. He wanted to be a Newton.

AN ENTIRELY UNKNOWN FORCE

By the end of November 1875 the investigations had reached a high pitch of excitement, and Edison was ready to make his conclusions, tentative though they were, public. He invited reporters into his workshop to interview him about the etheric force, resulting in a spate of newspaper articles in New York's *Sun*, *Herald*, *World*, *Tribune*, and *Times* that incited the public's fascination as well as incredulity. These articles revived a debate that had been going on for more than twenty years, focused on the validity of Baron Karl Ludwig Reichenbach's odic force. Reichenbach's theories, published in 1844, inspired new interest in America in 1851 when the first translation of his work appeared in English. Reichenbach had discovered what he called the odic, or magnetic, force in sensitive humans, and it seemed related to—even identical with—mesmeric powers. But as Reichenbach persisted in his investigations, he soon found that the force also emanated from crystals and could be identified by flames of light shooting from them. This discovery led him, in turn, to make grander propositions: the earth itself is odic, the aurora borealis is an odic light shooting from the planet, and proper alignment with the polarity of the earth could relieve humans of symptoms of illness. Sensitive people, for example, would do well to change the position in which they slept so that their head always pointed north. Reichenbach was discredited by the scientific community; when references to the odic force reappeared in articles about medical electricity, mesmerism, the current vogue for mind reading, and even reports of the research on psychical energies conducted by Sir William Crookes, an otherwise respected physicist, those references were disparaging. Still, as a popular theory, the odic force never really died, fueling a belief in a unified force of nature that linked nervous energy with magnetism, light, and electricity. Edison had been reading about Reichenbach just before he made his own discovery.[8]

While it was known that heat could be transformed into electricity, that electricity and magnetism were interchangeable, readers learned that Edison had discovered a new force that did not obey the same laws as heat, light, electricity, or magnetism. Beyond the wonders already known, something deeper and more complex existed in the universe. Some readers were thrilled: Edison, exulted a physician, "has touched the outer limits of one of the grandest discoveries of modern times."[9] But researchers reacted coolly. The articles were vague and misinformed, they complained, and they urged Edison to demonstrate his findings.

Preparing for such presentations involved tests to see how the force affected living tissue. Here, Beard's medical expertise was indispensable both to design the experiments and to affirm that there were no physical effects. "While we have the evidence of the sparks that the force is traversing the body," he testified, "yet, wherever directed, it causes no sensation, not even on the tip of the tongue, no muscular contraction anywhere, no tremor, no erection of the hair, no flashes of light, no sour taste, no dizziness—in short, none of the usual physiological reactions of the different forms of electricity."[10] In December, Edison's nephew Charley, who had joined his staff, became the primary test subject. Charley bravely hung on to a gas pipe with his feet above the floor, and, Edison's notebooks disclosed, "with a knife got a spark from the pipe he was hanging on." Charley stood on a block of paraffin eighteen inches square and six inches thick, and again elicited sparks from the pipe. Several muckers at once could draw equally strong and brilliant sparks, demonstrating that the force did not weaken as it passed through organisms, and when Edison himself grasped the gas pipe and held a piece of metal, he was able to obtain sparks from other metallic surfaces that he touched. The force clearly had passed through his body, and yet he felt nothing.[11]

These tests suggested that the new force might have significant commercial consequences: if it were produced from a strong apparatus, if it could be harnessed to yield either light or power, it might prove unimaginably safer than ordinary electricity. On the other hand, these initial investigations left open the possibility that the invisible, undetectable force was capable of causing insidious harm from long exposure, a suspicion that many already held about artificially generated electricity. What could the new force really do, over time, produced for widespread use?

Possible answers to that question might come from experiments with "the galvanoscopic frog," which were, Beard said, "the most delicate of all tests of electricity."[12] He offered to have his supplier deliver as many frogs as Edison might need, and Beard witnessed the experiments, which proved, perhaps not conclusively, that the etheric force was different from electricity. Sometimes the frog's muscle did seem to contract when the force was sent through it, but whether that contraction was caused by physical vibrations or atmospheric electricity had not been determined. Moreover, Beard admitted, no test had been made for chemical or thermal changes in the frog's tissue. Although Beard was ready to conclude tentatively that the force caused no physiological responses, Edison, writing to the editor of *Scientific American*, was unequivocal: the new force did not affect living matter.

Edison's staff plunged into other experiments, trying more than two dozen metals to determine which produced the strongest sparks. To see the sparks more easily, Edison and Batchelor designed what they called an etheroscope, a special dark box in which the spark could be observed clearly as it jumped a short gap. Concerned with reproducing his findings outside of his laboratory, Edison hoped that the etheroscope would travel to exhibitions around the world where scientists and ordinary visitors would be astounded by it.

While Edison worked to invent a new natural law, he also kept in mind that the force might have practical purposes. Some tests focused on its possible use in telegraphy, as a substitute for the wires that broke, fell, or electrocuted linemen. To that end, Edison designed a rudimentary experiment: "An uninsulated wire proceeding from the source of power (highly insulated, of course) was taken into the street and laid in the gutter around the whole block, and back into my laboratory by another door, and up to the floor above the one where the generator was. Excellent sparks were drawn from that end of the wire, although the ground the wire laid on was wet, it having rained all night."[13] If this test with uninsulated wires could be repeated and verified, then Edison believed he may well have found a cheaper, safer energy source for telegraphy: cheaper, because wires would need no costly coating; safer, because the etheric force did not cause shocks. For both reasons, the etheric force could mean a technological boon, but discovering a practical application for the new force was not Edison's main goal.

Edison continued his experiments, assisted by Beard, who worked for one intense week at Edison's laboratory and then pursued experiments on his own at the shop of a Manhattan telegrapher. Beard reported his experiments in a letter to the *New York Tribune*, which, as Jarvis B. Edson had suggested about the inductorium, proved to be a valuable endorsement. In that article and again in *Scientific American* in late January and in his own journal, *Archives of Electrology and Neurology*, Beard asserted that this new force was a kind of radiant energy, "somewhere between light and heat on the one hand and magnetism and electricity on the other, with some of the features of all these forces."[14]

As the new force was debated in newspapers and magazines, some scientists took up the challenge of disproving Edison's speculations and deflating the show of hubris implied in the term "etheric." Years later, the British mathematician Oliver Heaviside reminded Heinrich Hertz of the "great noise" made over the etheric force; "it led to nothing," Heaviside wrote, "and caused some ribald fellow to remark that Mr. Edison was supposed to have the new force 'concealed on his person.'"[15] Still, if it were not a new form of energy, it did appear to have surprising properties. The question remained: What had Edison found? Beard, in his article expanded and published as an offprint in 1876, tried mightily to present a cautious yet persuasive argument that the force was new, different, and, at the moment, inexplicable. He took particular pains to refute the claims by Edwin Houston and Elihu Thomson, two Philadelphia science instructors who devised their own experiment, that the force was nothing different from magnetic induction of electricity, except in this case the current produced was instantaneously reversed. Those findings, published in the *Journal of the Franklin Institute*, were noted with respect and relief by many scientists who doubted Edison's assertions. The article, though, did not convince Beard. No one had explained sufficiently the difference between this force and electricity: lack of polarity, lack of effect on the body, and failure to register on any measuring device. "An electro-magnetic apparatus is a reservoir of many forms of force—galvanic and induced electricity of various orders, magnetism, statical electricity, light and heat," Beard argued. "A source so rich in forces might give us at least one more; it is possible that one more has been here discovered."[16]

Edison and Beard were so intent on discovering a new force, so intent on creating a scientific revolution, that they did not realize what they had, in fact, found. More than a decade later, Hertz demonstrated that a rapidly oscillating electrical current would send out waves of electromagnetic energy that traveled at the speed of light and behaved in other ways like light. These waves, for example, could be deflected by metal, just as light could be deflected by a mirror or prism. Hertz's discovery formed the basis for producing radio waves. When Edison learned of Hertz's experiments, he realized that he and Beard had stood at the threshold of wireless telegraphy and radio technology. "What has always puzzled me since," he said later, "is that I did not think of using the results of my experiments on 'etheric force' that I made in 1875. I have never been able to understand how I came to overlook them. If I had made use of my own work I should have had long-distance telegraphy."[17] But in 1875 Edison was not trying to enhance his reputation as an inventor; he was seeking the radiance of immortality by identifying a new variation of the imponderable ether. In a sense, his imagination had been caught in the early nineteenth century.

MAGUS

Publicity about the etheric force in newspapers and magazines contributed to changing Edison's image from clever young inventor to a man with an extraordinary relationship to nature. In the summer of 1879, *Harper's Weekly* presented a portrait of the Wizard of Menlo Park that would characterize Edison for the rest of the century. For this reporter, the Menlo Park facility itself was a magical place. Edison's workshop, lit by flickering gaslight, contained "an infinite number of bottles of various sizes," "a wilderness of wires," "fiercely smoking lamps," and mysterious shadows. In this "uncanny" setting, the writer noted, Edison worked alone into the night. "His eager countenance is lighted up by the yellow glare of the unsteady lamps, as he glances into a heavy old book lying there, while his broad shoulders keep out the gloom that lurks in all the corners and hides among the masses of machinery. He is a fit occupant for this weird scene; a midnight workman with supernal forces whose mysterious phenomena have taught men their largest idea of elemental power; a modern alchemist, who finds the philosopher's stone to be made of carbon, and with his magnetic wand changes every-day knowledge into the pure gold of new applications and original uses."[18]

This portrait of Edison as magus, a portrait that emerged from many other periodicals and newspapers at the time, both reflected and created public opinion.[19] If the world was to change because of "new applications and original uses," that change, the article implied, should be generated by a man in touch with mysterious phenomena and supernal, or supernatural, forces; a man who was justly awed by nature's power. When Edison first heard his own phonograph talk back to him, he told a reporter, he was "actually frightened and for a moment dropped the crank he was turning."[20] Association with the occult and transcendent implied a man who was unworldly and, therefore, pure at heart. Others who strived to change the world—greedy robber barons and scientists who professed materialism and agnosticism—were not to be trusted. Edison, on the other hand, admitted that he was as suspicious of science as any ordinary man or woman. "He accepts nothing upon authority alone," a writer for *Scribner's Monthly* reported, liked books with "romantic ideas or ingenious plots," and apparently believed that all nature's creations were possessed of mind, desire, and will. "It is one of his axioms," the reporter discovered, "that all substances have an intelligence proportioned to their wants. 'Else why,' he asks, 'will a potato-vine travel one hundred and fifty feet in a dark cellar, and rise, against the law of gravitation, to seek a ray of light?'" When Edison shouted into his phonograph and heard the machine's responses, it did seem to his admiring observer that there was "an elfish personality there which has its own views of things and must be considered in its feelings." As for his motivation, Edison worked not for fame, "to which he is good-humoredly indifferent," but because of his "burning spark of inventiveness, and that only."[21]

Despite Edison's yearning to join the constellation of great men of science, his profession of anti-intellectualism served to separate him in the public's mind from the scientific community of which they were suspicious. Although science contributed to the technological innovations that the public might or might not embrace, might or might not find threatening, scientific premises seemed more dangerous than those technological innovations. "Fondness for the word 'scientist' is one of the notes by which you may know its votaries," William James remarked, "and its short way of killing any opinion that it disbelieves in is to call it 'unscientific.'"[22] Faith was one of those "opinions" that scientists and their votaries sought to kill. Science and technology were different: science appeared to deny the supernatural, arrogantly striving to explain the inexplicable; technology created ingenious inventions. For a while, anyway, Edison's reputation soared as he distinguished himself from malevolent scientists.

That image changed. By the late 1880s, Edison's projects were so diverse that Menlo Park no longer could accommodate them. With grand plans for a new workshop, Edison bought fourteen acres in West Orange, New Jersey, to construct, he said, the "best equipped & largest Laboratory extant."[23] The compound, when it finally opened in December 1887, consisted of an imposing four-story main building, sixty feet wide by two hundred long, housing Edison's library, offices, and personal experiment room; and four additional buildings devoted to chemical and physical laboratories, a chemical storeroom, and a metallurgical laboratory. A reporter who visited West Orange in 1889 described it not as a magician's lair lit by flickering flames but as a vibrant factory powered by "throbbing pulsating machines on each floor of the edifice." In his library, besides an abundant collection of scientific and technical works, Edison collected scrapbooks of "Edisoniana."[24] The size, the architecture, the scope of Edison's new laboratory reflected a man who had outgrown the image of a simple man among men, disdainful of experts, awed by the magic of his inventions. He was, reporters said, the "Napoleon of Science."[25] Edison had once characterized Napoleon, admiringly, as a hustler.

By the 1890s, Edison seemed less magician than mogul, as he involved himself in projects far more diverse than lighting systems and telegraphy. He scrapped ideas to invent artificial silk, an improved typewriter, and a flying machine, devoting his money and energy instead to such potentially lucrative enterprises as the separation of iron and gold from ore, the production of iron briquettes, and cement manufacturing; and to the development of the kinetoscope, motion pictures, storage batteries, electric cars, an improved phonograph and phonograph records, and poured-concrete housing. By the time he died in 1931, he had filed nearly 1,100 patents; received, among scores of other testimonials, a Congressional Medal of Honor; and counted among his friends Harvey Firestone and Henry Ford, with whom he occasionally went camping. No longer a lone genius, no longer a young man startled by his own talking machine, Edison had become, in the public's mind, an aggressive businessman and the world's most famous scientist. He was capable of thrusting unimaginable inventions into the marketplace, inventions that could generate unimaginable changes in the daily life of consumers of the force, and unimaginable wealth for the inventor. He seemed concerned with nothing more than the connection of power to money.

Beard gave up investigating the force, too, and turned his attention instead to expanding and elaborating upon the phenomenon of neurasthenia, its manifold cultural causes, and its connection to sexuality. The vital force, as Beard came to understand it, was far more complex than he had first believed. The flurry of excitement over the etheric force fueled interest—not only Beard's—in the connection among all energies and created expectations of new revelations about the universe and the bodies and spirits inhabiting it.

PART II

Cravings of the Heart

CHAPTER 6

THE INCONSTANT BATTERY

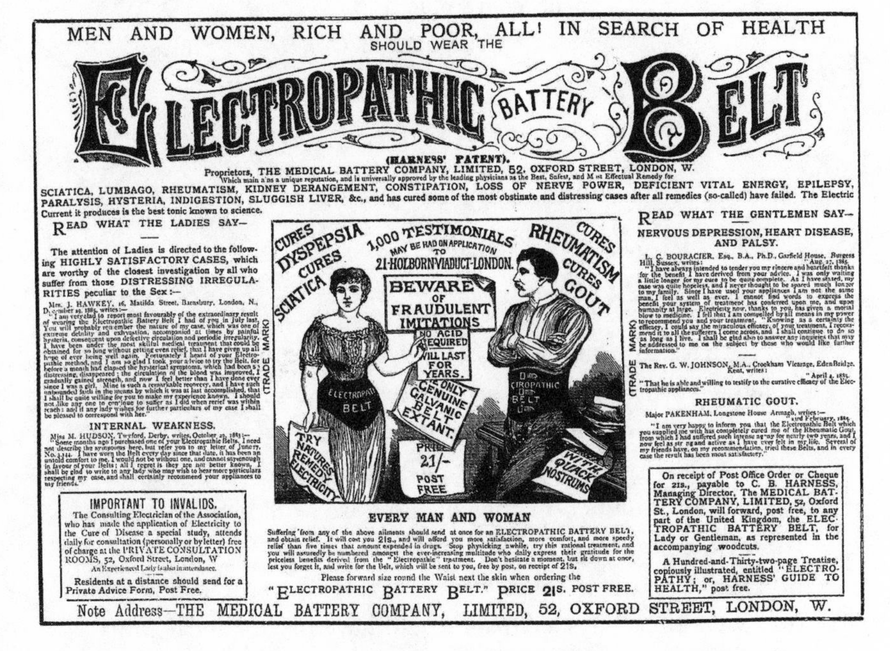

Advertisement from a British manufacturer for a popular electrotherapy device.

The Bakken Library, Minneapolis

All these forces with which we are so familiar—light, heat, electricity magnetism, motion, the vital force of plants, and the nervous force of man—are . . . forms of energy or power. They are all mutually convertible.

George Beard, *Our Home Physician*

From the late 1860s until the early 1900s, if you or I went to a physician complaining of persistent headaches, perhaps, or indigestion; depression or anxiety; paralysis (a frequently reported symptom of emotional stress); fatigue, insomnia, or night sweats; gynecological problems; pain in our legs, arms, or back; or any number of other ailments, we would have been offered electrotherapy, and we would have been disposed to try it.[1] The idea that energies were mutually convertible, that our own vital force was electrical, would have been familiar to us, not only from medical self-help books such as *Our Home Physician* but also from articles and fiction in newspapers and magazines. Certainly electricity seemed a better alternative than the commonly prescribed harsh drugs, alcohol-based tonics, or dietary recommendations that usually did not work. Electrotherapy often did work, although physicians themselves did not necessarily understand how: pain diminished, symptoms disappeared. Our friends and relatives would have told us about their experiences, and we would spread the word about ours. That is why electrotherapy became popular: partly because of the state of medical therapies at the time, partly because of widespread beliefs about vitalism.

Yet beliefs about vitalism that made people receptive to electrotherapy generated worry about other applications of electricity. Electricity simply should not be charging the very air that we breathe, exclaimed one Miss Caroline Smith in a piece entitled "A Prophetical Warning."[2] Electricity coursing through telegraph and telephone lines created a dangerous field causing "secret, subtle, and mysterious" effects on humans: apoplexy, palsy, paralysis of the brain, and nervous distress so profound that death would be a welcome remedy. At the same time that the public read news that electricity soon would pervade daily life, George Beard added electricity to his list of the causes of neurasthenia. While patients still accepted electrical infusions, they grew wary of the proliferation of the force. Electricity became a threat.

In the doctor's office, though, electricity seemed a wonder, not least for physicians themselves. Because their efforts to heal so often failed, because side effects of such drugs as arsenic, quinine, and strychnine were worse than the diseases they were meant to cure, many physicians were eager to find new ways to establish their authority and credibility. Electricity literally empowered them, offering an incontrovertible demonstration of expertise. Physicians who once armed themselves only with pills and potions took a new position in the treatment room: behind a large battery, manipulating wires and probes, immobilizing their patients, and sending a powerful force into their bodies. Electrotherapy conferred a new status on physicians as they aligned themselves with the rising young profession of engineering. "[I]n this electrical age," wrote William James Morton, professor of electrotherapeutics at the New York Post-Graduate Medical School, "the physician alone of educated men can not remain isolated from the general advance, can not lag behind in this great branch of scientific thought and practice."[3] Morton's statement, however, was not without controversy among physicians who believed that importing the findings of research science into medicine could undermine, not strengthen, the physician's authority and dehumanize the doctor-patient relationship. Clinical experience alone, some practitioners maintained, honed the intuition, responsiveness, flexibility, and imagination that were the hallmarks of the great physician.

In the mid-nineteenth century, science did not have a firm or necessary place in medical training, which was unregulated and inconsistent. Certainly George Beard's education was not typical of American medical students. One could enter medical school with no previous college preparation and complete coursework in two four-month semesters of lecture classes, with no experience in dissection and no hospital residency. Students heard lectures on inorganic chemistry, anatomy, the etiology and diagnosis of disease, surgery, and, sometimes, midwifery. Few medical students ever witnessed childbirth as part of their academic training. After this brief course of study, an aspiring physician would apprentice himself to a preceptor, usually a local doctor, reading the few books in his library and assisting with daily work. For this privilege, the apprentice paid the preceptor about $100 a year and gave him the benefit of free labor. Not surprisingly, preceptors were strong advocates of this system of medical training. After one year of lectures and two years of apprenticeship, the student was awarded a degree.

"A medical school is not a scientific school, except just so far as medicine itself is a science," Oliver Wendell Holmes told a Harvard medical class in 1867. "On the natural history side, medicine is a science; on the curative side, chiefly an art." Like many of his colleagues, he believed that practical experience served physicians more than scientific studies. Anatomy, physiology, and chemistry might offer some helpful principles but should not "crowd the more immediately practical branches... The bedside," he added, "is always the true centre of medical teaching."[4]

Some physicians, especially those who had been trained in France or Germany where laboratory science and hospital experience were part of the curriculum, protested for reform. But the lecture format made medical schools cheap to operate, and administrators were reluctant to change a system that had proved so profitable. In 1847, for example, when the University of Pennsylvania Medical School and New York's College of Physicians and Surgeons advertised their newly developed six-month course of lectures, enrollment plummeted. Alarmed, the schools reverted to their traditional four-month semesters.

Just as medical education was slow to reform, many physicians resisted changes in clinical practice, especially if those changes relied on new technology. Early in the nineteenth century, a medical examination consisted largely of observation: taking the pulse and observing skin color, manner of breathing, and appearance of urine. The stethoscope, invented in France in 1819, was only slowly accepted by physicians, who spurned the use of instruments. These artifices, they believed, would diminish, rather than enhance, their authority, because physicians would appear to be nothing more than surgeons, then at the lowest rung of the medical hierarchy. Hermann von Helmholtz, looking back on his own medical education, remarked that to those who trained him, using instruments seemed to be "a coarse mechanical means of investigation which a physician with a clear mental vision did not need; and it indeed lowered and debased the patient, who was anyhow a human being, by treating him as a machine. To feel the pulse seemed the most direct method of learning the mode of action of the vital force, and it was practiced, therefore, as by far the most important means of investigation. There was, as yet, no idea of measuring temperature in cases of disease." Helmholtz became more aware of these attitudes after he invented the ophthalmoscope, which allowed physicians to examine the retina. One colleague told him that "he would never use the instrument, it was too dangerous to admit crude light into diseased eyes; another said the mirror might be useful for physicians with bad eyes, his, however, were good, and he did not need it."[5]

Yet in time physicians discovered that the stethoscope gave them an intimate, and secret, relationship to the patient's body, even though convention dictated that they perform auscultation through the underclothes. Instead of getting information about symptoms from a patient's narrative, now the physician listened in silence through a rigid metal tube to sounds that the patient could neither hear nor interpret. The body's interior became a site outside of the patient's expertise.

Despite the stethoscope, laryngoscope, speculum, and ophthalmoscope, most physicians realized that they could do little for the most prevalent infectious diseases—pneumonia, typhoid, tuberculosis, and diphtheria—nor for recurrent outbreaks of cholera, smallpox, or influenza. In 1855, Jacob Bigelow, chair of materia medica at the Massachusetts Medical College, proposed to his colleagues that many diseases were "self-limited" by their own course and not able to be affected by "foreign influences," such as drugs prescribed by physicians.

Among those illnesses were such common afflictions as measles, scarlet fever, whooping cough, and mumps.[6] Nevertheless, patients expected treatment and wanted to believe that their physician's intervention could cure.

Medical therapeutics focused on treating symptoms, rather than specific diseases, with drugs that included antipyretics, such as quinine, to reduce fever; analgesics or hypnotics, notably opium, to relieve pain; cathartics and emetics to rid the body of excess matter; and tonics to stimulate the pulse and digestion. Besides harsh drugs with odious side effects—mercury, arsenic, and antimony—physicians relied on physical treatments such as bloodletting, cupping, blistering, and leeches. Every doctor's office, one physician recalled, contained "a desk, a few hard chairs, a tall closet containing an entire articulated skeleton...and the lancet and a jar of leeches were always at hand."[7]

For many illnesses, physicians believed, health would be restored if bodily fluids were properly regulated by extracting blood or promoting perspiration, urination, defecation, or salivation. These processes often caused additional suffering. Mercury, for example, widely used to treat syphilis and as the purgative calomel, produced excessive salivation and soreness of the gums even when taken sparingly; after prolonged use, it caused mouth ulcers, loss of teeth, and eventually destruction of the facial bones. It is no wonder that physicians called these treatments "heroic."

Robert Koch's discovery of the tuberculosis bacillus in 1882 generated excitement and dissent in the medical community. Could it be, some physicians proposed, that the bacillus was a result of the disease and not the cause; could it be that it appeared in the body coincidentally with tuberculosis? As late as 1888, when the journal *Science* sent a questionnaire to every

American medical school asking about their faculty's belief in the theory that most infectious diseases are caused by microscopic organisms, responses were not unanimous. The Minnesota College of Physicians and Surgeons expressed doubts, as did the American Medical College of St. Louis, Long Island College Hospital, and the Hospital College of Medicine in Louisville, Kentucky. The teaching of bacteriology varied widely, with some schools having a fully equipped laboratory and special lecturer, while others, such as the New York Medical College for Women, assigned a professor of hygiene to teach "something of the theory of germ-cells and microbes in disease, and the importance of care and cleanliness."[8]

The theory that illness could be caused by invasive microscopic agents supported the idea that electricity also could be an invasive power. "The 'seeds of disease,' to adopt a popular term (whether we accept the *germ theory* or not)," commented *Scientific American*, "are floating about us in myriads without number, and are inhaled by us with every breath, and yet the diseases are manifested only here and there, wherever the 'seed' finds a susceptible point for its growth. In the same manner, though the electrical influence may come alike upon all, yet is its effect made manifest to us in certain cases with great power, while in others we fail to detect it." Physicians noticed that some people were more sensitive to electrical changes than others, reporting tingling of the skin, for example, hours before a thundershower.[9] In these people, electricity could incite nervous symptoms and result in neurasthenia.[10]

Competing theories about the causes of illness led to a wide range of therapies. Patients sought those, like homeopathy and hydropathy, in which treatments were not painful, and the patient, therefore, did not feel punished for being ill. Of these alternatives, homeopathy attracted an enthusiastic following, partly because it was based on the premise that the body could be stimulated to heal itself, and partly because homeopaths allowed patients more authority in the medical partnership than did conventional therapists.

Homeopathy was developed by German physician Samuel Hahnemann in the late eighteenth century. Dissatisfied with the heroic therapies available to him, Hahnemann experimented on himself by administering a variety of medicines to test their effect on a healthy body. He discovered, much to his amazement, that certain drugs induced symptoms of certain diseases, and he concluded that these drugs would prove most efficacious in curing those very illnesses. A fever-inducing drug would be prescribed for fever, purgatives for diarrhea: like cures should be given for like illnesses, he maintained, because illness had no biological or anatomical cause, but resulted from a disturbance in the patient's vital force. Agents that afflicted the patient's system with more intense symptoms would incite a resurgence of the vital force needed to expel the disease. Hahnemann modified the dosage of these symptom-inducing agents until he found the smallest possible amount needed to restore vitality, and he sometimes incorporated mesmerism into his therapy, believing that mesmeric passes were helpful in redistributing the vital force throughout the body.

Like hydropathy, which sent patients to spas where they bathed in and drank allegedly curative waters, homeopathy seemed to be more "natural" than conventional therapeutics. Homeopaths paid close attention to the delicate shadings of symptoms, listening attentively to patients' own renderings of their feelings and responses, and they treated patients with a more imaginative and extensive arsenal of substances than conventional physicians. The therapy proved popular in France and, by the 1840s, in America, where the American Institute of

Homeopathy and the Homeopathic Medical College gave practitioners a certain amount of credibility. Nathaniel Hawthorne, William Wadsworth Longfellow, Louisa May Alcott, Henry James Sr., and, at times, other members of the James family were followers of homeopathy, all eager for the kind of authority that homeopathy gave to the patient; all convinced that homeopathy worked to shore up the body's strengths rather than to deplete vitality, as did bleeding or purging; all skeptical of conventional practitioners. Patients attracted to homeopathy or hydropathy were inclined to try electrotherapy, too, which they believed literally empowered them. As they became convinced that electrification could cause illness, electrotherapy became more and more appealing.

EXCITING CAUSES

Soon after Beard published his first paper on neurasthenia in 1869, he began to elaborate on the diagnosis, expanding the range of symptoms that could be identified as neurasthenic and drawing increasingly on technological metaphors to explain the illness to colleagues and patients. For Beard, light, for example, served not as metaphor of moral triumph, but as a symptom of fragile nervous force. Beard compared the neurasthenic patient to "an electric light attached to a small dynamo and feeble storage apparatus, that often flickers and speedily weakens when the dynamo ceases to move." The neurasthenic patient suffered from having "a battery with small cells and little potential force... an inconstant battery, evolving a force sometimes weak, sometimes strong, and requiring frequent repairing and refilling." These empowering mechanisms could be "broken or eaten away by chemical changes" and, like one of Edison's dynamos, needed vigilant attention to be kept in working order.[11]

Striving for wealth, competitiveness, and family pressures were among the "special exciting causes" of the illness, but now Beard recognized other stresses, some peculiar to American culture. Americans, he found, were more sensitive than their European counterparts to stimulants such as wine and cigars; they ate faster and drank less fluids than Europeans; their physical constitution was more delicate, making them more susceptible to such debilities as tooth decay and hay fever; and they had to meet the demands of living in a democracy, which gave everyone monarchical responsibilities and consequent stresses. In addition, Beard identified a few recent cultural changes as precipitating causes: "steam power, the periodical press, the telegraph, the sciences, and the mental activity of women."[12]

Steam power and the telegraph most directly shaped his patients' perception of the rapid pace of life, which had been implicated in Beard's original diagnosis of neurasthenia ten years earlier. Before the telegraph, international business dealings were enacted at a leisurely pace, with both messages and cargo transported by ships. Now telegraphs communicated information, such as price fluctuations, so quickly that businessmen felt tyrannized by the "constant knowledge," and the pressure to act decisively in an atmosphere of intense competition. Steam power also moved business more quickly: new machinery increased the quantity of work that could be done, and it also changed the nature of that work. Instead of being artisans, laborers found themselves required to do repetitive and boring tasks, "depressing," Beard noted, "both to mind and body." The increase in nervous diseases among laborers was a direct result of these "unanticipated" effects of steam power.[13]

Identifying the mental activity of women as a cause of neurasthenia suggests that Beard was treating, or hearing about, an increased number of female patients, some of whom were new members of the workforce. Between 1880 and 1890, the number of female workers nearly doubled, to four million, mostly in jobs created by new technology: telegraph and telephone operators, typists, and incandescent-bulb makers. Although women were hired for some jobs more for their small and nimble fingers than for their intellectual capacity, Beard recognized the stresses inherent simply in being employed. From the first, neurasthenia had been an illness most often affecting male brain workers; but now women, either in the workforce or at home, were becoming exhausted, depressed, and anxious enough to confide their symptoms to their physicians. And physicians like Beard differed from colleagues by explaining women's nervous symptoms not necessarily in relation to sexual organs, but in relation to intellectual labor and cultural stresses, such as newspapers and magazines.

The periodical press, Beard realized, could powerfully manipulate a reader's perception of reality. The press surely offered his patients more opportunities for tiring brain work, surely provided so much information that his patients might feel overloaded. Even intelligent readers, Beard said, felt frustrated when they read articles about Edison, confronted with the "multitudinous directions and details of the labors of this one young man with all his thousands and thousands of experiments and hundreds of patents."[14] But more subtly, the press created an atmosphere of impatient anticipation. It thrived on and promoted a world of exciting, ever-changing events, reporting these events as if they had direct and immediate impact on readers, whether they did or not. Coverage of Edison's invention of the incandescent lamp was ample evidence of the way the press created an escalation of hope, a plummeting of disappointment and betrayal, and an overwhelming sense of urgency.

In his menu of causes, though, science seems an odd entry: Beard was known to be a champion of science as a discipline and scientific method as a way of knowing the world. Yet he recognized that science created a culture of expertise which undermined people's belief in their own authority. Instead of trusting what they could see and hear, people were persuaded that their senses could deceive them. Only scientists, with their knowledge of theory and special access to invisible natural phenomena, were able to distinguish true from false. This undermining of personal authority led to feelings of anxiety and powerlessness, two symptoms of neurasthenia.

THRILLS IN THE NERVES

Along with an expanded list of the possible causes of neurasthenia, Beard turned his attention to a special manifestation of the illness in men. In the 1880s, he revised his diagnosis in a bold new direction, responding to recurring complaints by male patients of "genital debility." This special weakness of the vital force was so prevalent that Beard decided it required special consideration. Women's neurasthenia, as Beard knew, frequently manifested itself as gynecological symptoms, usually treated successfully with electrotherapy. Physicians reported such cases in medical journals, of which Mrs. B., a thirty-seven-year-old mother of six, was typical. Married for sixteen years, Mrs. B. complained to her physician, Boston doctor Edward Reynolds, of "a dragging sensation, great nervousness, unexplained crying spells, inability to leave the house alone." Reynolds, rather than prescribe bromides or a rest cure, concurred with his fellow electrotherapists that Mrs. B. could get better tonic and sedative effects from electricity. Since many physicians believed that disturbances in the reproductive system incited, by reflex irritation, symptoms in other parts of the body, they felt justified in administering electrotherapy to the genitals to alleviate other complaints. Mrs. B.'s crying spells, her agoraphobia, all of her symptoms might be related to weakness in her uterus. Reynolds decided that the best treatment would be faradic currents applied through the vagina for ten minutes every week. The sensation, Mrs. B. found, was not at all unpleasant. Treatments continued for four months, after which, Mrs. B. reported, she was able to go shopping in New York with her husband. "Something," she said, "I never expected to do again."[15]

Physicians lauded electrotherapy for gynecological complaints, although they were puzzled about why the treatment should work for opposite conditions—lack of menstruation in one case, profuse bleeding in another. Nevertheless, they were pleased by their patients' reactions to "a slightly hot tingling," as one patient put it, or "a little well pain." Moreover, electrotherapy insured a patient's modesty: the electrodes could be inserted "over the pubes" or even into the vagina without "any undressing of the patient, or at most only the removal of the drawers."[16] If gynecological symptoms occurred in neurasthenic patients, when those symptoms improved, physicians found that other complaints—backache, headache, insomnia, depression—disappeared or at least diminished.

Beard saw, however, that physicians responded differently to neurasthenia in men. When men complained of depression, "morbid fear in all its types and phases," palpitations, "deficient mental control," many physicians, even years after Beard had publicized his diagnosis of neurasthenia, were likely to dismiss them as hypochondriacs. Beard meant to correct that mistake by offering a new variety of neurasthenia applicable mostly to men: sexual neurasthenia. American men living in the north-east were especially susceptible, partly because of the extremes of climate, partly because of the "opportunities and necessities of a rising civilization in a new and immense continent." Because Beard subscribed to the theory of reflex irritation, he did not take the prevalent view that genital debility was caused solely by excessive sexual activity or masturbation; too much brain work could bring on the symptoms, as could too vigorous competition for wealth. What all men needed was a tonic in the form of electricity. "We speak of toning a violin so that its strings give forth a clearer and a better sound," he explained. "This we do by making them more tense." In the same way, tonic therapies "gradually produce the requisite degree of tension of the nervous system, and of the living fibre generally, and which enable it fitly to respond to all its natural and appropriate stimuli."[17] Electricity enhanced a sinewy, vigorous image of manhood.

Beard's prescription of electricity as tonic for sexual problems was far different from other physicians' use of electricity as a form of punishment for what they considered to be sexual excesses. Painful shocks, they believed, would act as a deterrent. Physicians fitted some patients with an "electric monitor," designed to emit a strong shock when the wearer had an erection.[18] They applied electricity through the rectum or urethra, on the spine, or directly on the penis to produce far more than the "slightly hot tingling" experienced by their women patients. Underlying their treatment was a belief that loss of seminal fluid and excessive sexual excitement resulted in loss of nerve force. Beard, by connecting sexuality to mental health and cultural stresses to neurasthenia, liberated men from condemnation. He himself had suffered recurrent neurasthenia until he met Elizabeth Ann Alden and, in 1863, asked her to marry him. Beard adored Lizzie, thought about her every moment of the day, dreamed about her at night. "I breathe Lizzie," he exulted in his journals during their courtship.[19] When his partnership with Alphonso Rockwell finally insured him an income, he and Lizzie married—like the Edisons, on a Christmas Day. Beard's letters to Lizzie, on the occasions when they were apart, attest to an enduring love. But a happy marriage was not the only cure: Beard also overcame neurasthenia by rejecting the repression inflicted on him as a youth and by modifying what his culture taught him about self-control. His self-prescribed asceticism gave way to temperance, then finally to exuberant sensual indulgence in food, drink—including alcohol and coffee—music, and dancing. He liked to lunch well at Delmonico's and believed that rare roast beef was an antidote to depression. "Meat," he proclaimed, "was the soul of the neurotic diet," and animals closest to humans would provide the best nutrition: monkeys, for example, rather than cows.[20] His lack of temperance stretched to money; he was frequently in debt and hounded by creditors, but he chose pleasure, however precarious his finances, over a pinched life.

Personal disposition, though, only partly explains Beard's new perception of illness. Just as his original diagnosis served to build up his practice, the diagnosis of sexual neurasthenia offered physicians a chance to broaden their clientele. By the 1880s, many patients had taken electrotherapy into their own hands. Electric hairbrushes, teething rings, back supports, and amulets: all these and more were available to individuals by mail order from advertisements in newspapers and magazines and through such venues as the Sears Roebuck catalogue. Although physicians protested against what they deemed as quackery, they realized that the advertisements reflected the public's needs and desires. And what the public desired, they could not fail to notice, was treatment for sexual problems.

Many popular devices were designed to administer pleasurable doses of electricity to the genital area for both men and women. "WEAK MEN!" one advertisement proclaimed. "Debilitated through Indiscretions or Excesses, We Guarantee to cure by this New Improved Electric belt & Suspensory."[21] For $1.50, a consumer could buy Dr. Tuckett's forty-cell electric belt, advertised to cure "grip, rheumatism, lost vitality."[22] Women could buy corsets illustrated with sparks of energy emanating below the waist. "Exhilarating!" one ad proclaimed. "The 'very thing' for Ladies," boasted another. The Medical Battery Company of London featured the Electropathic Battery Belt, worn like an apron on top of a woman's underclothes, promising to cure "Distressing Irregularities peculiar to the Sex." Those irregularities included exhaustion and hysteria, which according to one patient's testimony disappeared marvelously after a month of wearing the belt. It seemed in physicians' interest to name these symptoms and publicize their own expertise in treating them.

As Beard was writing about sexual neurasthenia, however, he doubted that some colleagues would greet this diagnosis with any more enthusiasm than they had given to his first paper on neurasthenia more than a decade earlier. Even as late as 1884, Roberts Bartholow, a prominent Philadelphia physician whose electrotherapy manual rivaled Beard and Rockwell's, rejected the idea that neurasthenia was a "primary nervous affection, or . . . a substantive disease. I hold," he insisted, "that it is symptomatic and secondary." Although patients complained of depression, anxiety, blurred vision, dizziness, fear of crowds, indecisiveness, or impaired memory, Bartholow believed that these symptoms resulted from organic causes and could be treated by change in diet or work habits, as well as by electrotherapy. Only when somatic illness existed could stress from too much work or civilization lead to emotional or psychological symptoms. Nor did Bartholow believe that Americans had any special tendency to neurasthenia. It might be called the French disease, he suggested, since France was subject to even more political unrest than America and surely boasted as much urban excitement. He worried that by publicizing neurasthenia, Beard had given patients permission to study "bodily sensations" too closely. "All the world knows that when the attention is strongly fixed on an organ of the body functional disturbances of it ensue, and finally structural changes may be induced." It might be, he suggested, that many symptoms of neurasthenia were caused by reflex action in the brain, affecting the heart, the eyes, the stomach, the muscles, and the genitals.[23]

Beard anticipated more sympathy for his innovative diagnosis from the German medical community, where his previous work had been translated, quoted, built upon, and heralded. When Beard compared the medical profession in Europe and America, he found little good to say about his native land. In England and America, he asserted, physicians received the poorest education, since an undergraduate degree was not required for medical school. This lack of a liberal or scientific education made physicians provincial and unimaginative. "With us who speak the English language," he wrote, "the presumption is, that a man who originates a new idea is insane." While America and England suffered from "mental myopia," Germany valued novelty. "When an idea appears on the horizon the German asks is it true; the Englishman, is it safe; the Frenchman, was it born in France; the American, what does Europe think?"[24]

Europe thought well of neurasthenia, although physicians demurred about its being a particularly American disease. There were neurasthenics in England, France, Italy, and, most definitely, in Germany and Austria. Sigmund Freud called neurasthenia "the commonest of all diseases in our society," and he knew Beard's work well.[25] But he believed that Beard did not go far enough in his diagnosis and certainly not in suggesting an etiology for the illness. The pressures of civilization simply did not suffice. As Freud considered neurasthenia in his patients, he focused less on his patients' symptom of exhaustion than on what Beard had characterized as "morbid thoughts" and Edward Van Deusen, in his early essay on neurasthenia, as "distrust," or lack of faith in whatever beliefs—religion or love, for example—had sustained the patient before the onset of illness. Neurasthenic patients, in Freud's view, felt disillusioned, betrayed, and powerless to control their lives. This powerlessness Freud called "the anxiety psychosis," which, he said, "is never missing and, whether denied or admitted, betrays itself by a profusion of newly emerging sensations, that is, by paresthesias."[26] Besides manifesting itself in annoying sensations of prickling, tingling, or itching, anxiety resulted in depression, various phobias, pessimism, and "a certain lowering of self-assurance." Anxiety might occur in sporadic attacks—common in women—or as a chronic state, most often in men. Heredity might predispose someone to anxiety, but it was not the ultimate cause, nor did heredity explain why men suffered more chronic anxiety than women. If there was a gender difference in the symptoms, it seemed likely that there was a gender difference in the cause. This difference, Freud decided, could be found in sexual experience. "No neurasthenia or analogous neurosis," Freud claimed, "exists without a disturbance of the sexual function." This disturbance might take the form of masturbation, coitus interruptus, or "sexual traumas before the age of understanding begins."[27] Although Freud retained the term *neurasthenia*, nerve weakness, in effect he redefined the term to mean loss of sexual vitality or, as it had been called in earlier times, animal electricity.

Beard, however, never knew of Freud's reconsideration of his work. On Wednesday, January 17, 1883, he visited a dentist complaining of an ulcerated tooth. After the tooth was extracted, however, Beard's symptoms did not improve and in fact worsened. At first his physicians suspected blood poisoning; but on Sunday, visiting his bedside at the Grand Hotel in Manhattan, they diagnosed pneumonia. Beard realized that his condition was dire, but although he suffered considerable pain, he retained, his friends said, the cheerfulness and inquisitiveness that always had been his most remarkable traits. He wished, he told them as he lay dying, that he could make a record of these final profound struggles; he wished, he repeated often in those last days, that others would continue the work in psychology and medicine that he felt he had only just begun. He died on Tuesday, January 23, at ten thirty in the morning; he was forty-four. A postmortem examination confirmed that "the absorption of purulent matter of the upper jaw" had caused, or perhaps accelerated, embolic pneumonia. Funeral services were held on January 27 at the Broadway Tabernacle, and Beard was buried in Andover, Massachusetts. The church was cold and damp on that January day, and Elizabeth Beard, within days, was ill with pneumonia. A week later, their eight-year-old daughter, Grace, was an orphan.

In the fall, eulogizing his former partner at the American Academy of Medicine, Alphonso Rockwell evoked Beard's egotism, his "sense of placing a high value on his own interpretation of certain phenomena in physics and psychology." As Beard had often said of himself, Rockwell reminded his listeners, "he never argued, he simply asserted."[28] Although their partnership had ended years earlier when Beard decided to devote himself to writing about nervous illness, spiritualism, the value of expert testimony, and assorted other controversial issues, Rockwell never wavered in his belief that neurasthenia was an important diagnosis and electrotherapy an effective treatment for that and other ailments. He was awarded a professorship of electrotherapeutics at New York Post-Graduate Medical School and Hospital and emerged as a respected medical consultant. After Beard's death, he worked on shaping Beard's last manuscript for publication: *Sexual Neurasthenia* appeared in 1884. By 1898, it had gone into its fifth edition.

ELECTRICAL LADIES AND GENTLEMEN

The conflation of sexual vitality and animal spirits that Beard and Freud asserted was not limited to the physician's examining room but emerged in popular fiction as testimony to the enduring belief that electricity was the essence of life. Hamlin Garland, a well-regarded journalist and fiction writer, published "The Electric Lady," for example, featuring a traveling performer renowned for having "the greatest amount of animal magnetism" ever seen. She tours the country, amazing her audience by first lifting three men seated on a chair and then by defying any man to lift her—an impossible feat, it turns out, until her elbows are wrapped in silk handkerchiefs, which obstruct the colossal force of magnetism holding her firmly on the ground. But her effect on the men in her audience has less to do with her performance than with her bold sexuality: she stands before them "lithe and muscular in her fine, rich, close-fitting dress, which came to her knees . . . Her whole bearing was firm and deft and confident." The story turns on her relationship with Sam, a virile young man who pits his own "crushing power" against her "slender body with its amazing resistance"—resistance that offends his "physical pride." While the woman is able to resist Sam, he finds her irresistible and vows to follow her to the city where she lives, and where, he imagines, his life will change completely. He plans his trip "with a magnificent feeling of being his own master" in what would be "the most extraordinary experience of his life." Sam would achieve this sense of authority over his life in part because he is daring enough to leave the rural town where he is a farmer, bound to the routine of work and church and Sunday drives with a local girl. But part of his new feeling of mastery comes from his acceptance of electricity and all that it implies: urbanization, new stimuli, and sexual excitement. Sam never does manage to leave home; but at times he remembers the electric lady and her power over him. "It was the greatest romance," he thinks, "that had ever come into his life."[29] The romance between Sam and the electric lady was a prototype of other more subtle fictional love affairs, where the electrical spark between a man and a woman signified sexual attraction.

The connection of electricity to sexuality and vitality was blatant in Scottish writer James Maclaren Cobban's thriller, *The Master of His Fate*. Cobban's protagonist, the physician Lefevre, believes, like many of his contemporaries, that the "universal principle in Nature" is electricity, which could be controlled so as to yield "a universal basis of cure."[30] When two patients, male and female, are brought to his hospital in a comatose state, he revives both by infusing them with electricity generated by a machine of his own invention, a machine of which his body is an integral part. After attaching conductive wires to the patients' hands, Lefevre places his fingers in the chemical bath of the battery and experiences a "convulsive shudder," not unpleasant, as electricity surges from the battery, through the wires, to his patients.[31] These two patients, however, are unlike others he has treated: they are unconscious not because they have been hypnotized irresponsibly (an occurrence not rare in nineteenth-century Britain, and sometimes offered as a defense for crimes), nor are they suffering from hysteria, which often results in swooning. Instead, they recount the same mysterious tale: they met a strange old man who enthralled them and then laid his hand upon them. His touch felt thrilling—as if, the male patient admitted, a woman had touched him—and that touch was the last thing they remembered.

As he investigates the mystery, Lefevre discovers that the stranger is none other than his enigmatic friend Julius, who, like a technologically updated vampire, must draw his life force from the animal magnetism, the vital energy, the nervous ether of others. Electricity, Julius proclaims, is the spirit of life; it "flows and thrills in the nerves of men and women, animals, and plants, throughout the whole of Nature! It connects the whole round of the Cosmos by one glowing, teasing, agonizing principle of being, and makes us beasts and trees and flowers all kindred!"[32] But Nature's bounty is not limitless, as Julius well knows; "all activity," he explains, "all the pleasant palpitation and titillation in the life of Nature and of Man, merely means that one living thing is feeding upon or is feeding another."[33]

In his youth, Julius manages to derive electrical nourishment from Nature—caressing small animals, for example—but eventually he must prey on young, healthy humans, male as well as female. His curious debility convinces him that all living things are "animated by one identical Energy of Spirit of Life" and "have the power of communicating, of giving or taking, this invisible force of life." The channels of communication, he tells Lefevre, are through nerves, "so that as soon as a nerve in any one shape of life touches a nerve in any other, there is an instant tendency to establish in them a common level of the Force of Life." When someone clasps a friend's hand, for example, or embraces a lover, "the force seeks to flow... from one to another, evermore seeking to find a common level,—always that is, in the direction of the greater need, or the greater capacity."[34] The disparity between most people is small, making this fluctuation of energy imperceptible. But voracious humans like Julius are condemned to live upon, to consume, to deplete others. Julius's desperate need inspires in Lefevre a sudden medical insight: it is one thing to create a machine to revive an unconscious patient by "stimulation of Will and Electricity." Why could he not create a means—an enhanced form of mesmerism, actually—to revitalize from a source of human energy? "Why," he asks himself, "should there not be Transfusion of Nervous Force, Ether, or electricity, just as there is Transfusion of Blood?"[35]

The idea is titillating: infusing the nerve force of another human being means achieving a rare intimacy. The nervous ether, Lefevre explains, "is clearly very volatile, and at the same time a very searching fluid. It can easily pass through the skin from a nerve in one person to a nerve in another. There is no difficulty about that; the difficulty is to set up a rapid enough vibration to whirl the current through!" Lefevre does invent this device, using an electrified tuning fork and a violin bow to turn his body into an instrument capable of "searching" into the body of a female patient. The effect is thrilling, even orgasmic: "The vibratory influence whirled wildly though him, there was a pause of a second or two... and then suddenly a kind of rigor passed upon the form and features of his patient, as if each individual nerve and muscle were being threaded with quick wire, a sharp rush of breath filled her chest, and she opened her eyes and then closed them again."[36] Lefevre, though, is left exhausted, his own nerve strength depleted.

The Master of His Fate, published two years after Robert Louis Stevenson's *Dr. Jekyll and Mr. Hyde*, is more than a study of dual personality and more than a variation on popular vampire stories. Reflecting the apprehensions of the time, the novel, with its ironic title, celebrates the mysterious power of electricity at the same time as it offers a dark, cautionary tale about transgression—sexual and intellectual. As a physician, Lefevre rejects scientific materialism, erring, his colleagues believe, too much toward mysticism. His travels to India and the Middle East have offered him glimpses of alternative therapies, and he returns believing that Western medicine is "inexact and empirical... based merely on custom, and a narrow range of experience." Yet his daring to defy those customs gives him unseemly power over his patients and threatens his own existence. Lefevre is eager to revise his beliefs in natural law, but when those beliefs are toppled, Cobban warns, one's sense of stability, psychological integrity, and power also may collapse: "Have you ever found something happen or appear," Julius asks, "that completely upsets your point of view, and tumbles down your scheme of life, like a stick thrust between your legs when you are running?" Humans can delight in Nature, but if they seek to control it, they do so at their peril. "Outraged nature," Julius declares, "exacts a severe retribution!"[37]

CHAPTER 7

Haunted Brains

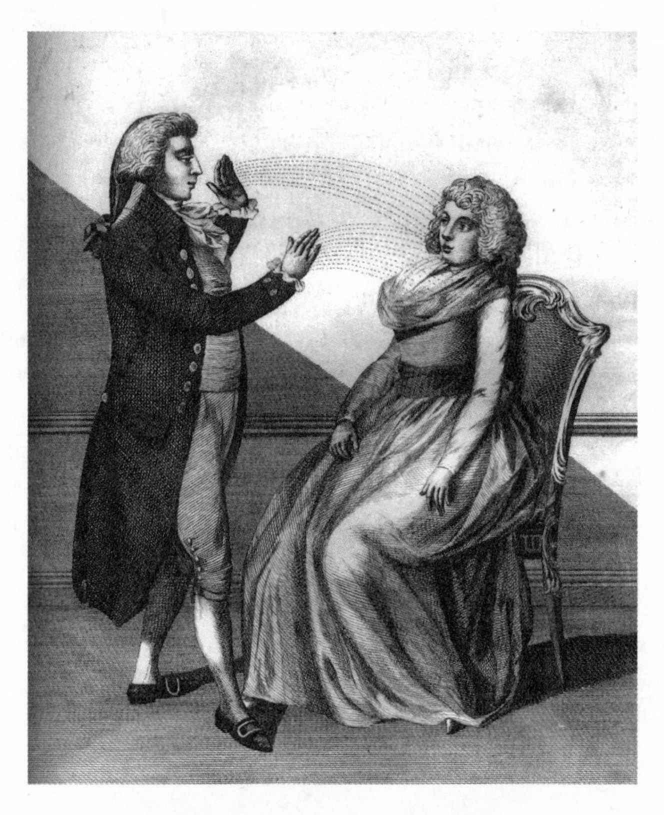

Mesmerism, from E. Sibley's A key to physic and the occult sciences *(1794). Magnetic emanations from the mesmerist's hands create the subject's highly emotional state or "crisis."*
The Bakken Library, Minneapolis

For my part, I am inclined to believe that our personality hereafter will be able to affect matter.

Thomas Edison

When Walt Whitman sang "the body electric" in the 1850s, he was not being metaphorical. Like his contemporaries, he believed that electricity animated the body; and not only the body, but the mind, too. Emotions, thought, will, and spirit—all were manifestations of electrical energy. This energy, radiating mysteriously, made it possible for one person's will to influence another's; made it credible that the soul persisted, somewhere, after the body's death. Spiritualism, promoting these ideas, fascinated the public from the middle to the end of the nineteenth century. You or I might well have been among that public. For 25¢ we could buy a ticket for a spicy demonstration of mesmerism, hypnotism, or mind reading. We could listen to any of thousands of traveling lecturers speaking about "phreno-magnetism" or "nervo-electric influence." Universalist minister John Bovee Dods, prominent among them, lectured on mesmerism before the United States Senate, by invitation of Daniel Webster, Sam Houston, and Henry Clay. Physician Robert Hanham Collyer, who toured internationally, published his talks as *Mysteries of the vital element in connexion with dreams, somnambulism, trance, vital photography, faith and will, anaesthesia, nervous congestion and creative function.*[1] We might have bought that book, or hundreds of others. We could join our friends at séances, eager for messages from the dead. Souls, we would believe, vibrated with a mysterious electric charge that could be detected by mediums. Each of us emanated electricity, and we could be aroused sexually and emotionally by electrical currents generated by others.

As with beliefs about vitalism, spiritualism helped to fuel resistance to electrification. If the force could invade the mind, it needed to be controlled, not let loose in the world. As electricity was being marketed for mass consumption, the public's interest turned enthusiastically to spiritualism.

Mesmerism and hypnotism fired the public's imagination. Both suggested a connection to the realm of the supernatural, but they had another appeal as well. Both served as a path to self-knowledge; both served as medical therapy. Many who submitted to mesmerism or hypnotism were suffering the effects of mind upon body to generate pain; they sought to understand those effects and, somehow, to control them. In June 1844, for example, Harriet Martineau agreed to try mesmerism to cure what had been diagnosed as uterine cancer and to release her from opium addiction. In severe pain for six years, debilitated, and constantly nauseated, she was desperate for help. Although she was initially skeptical of the controversial therapy, within her own circle of friends, many had tried mesmerism and applauded it. Faced with medicine's "dense ignorance of the structure and function of the nervous system," Martineau concluded, why would any patient refuse treatment that promised relief? "Whatever quackery and imposture may be connected with it, however its pretensions may be falsified, it seems impossible," she said, "but that some new insight must be obtained by its means, into the powers of our mysterious frame."[2]

By 1844, mesmerism and hypnotism had become interchangeable terms, conflated sometimes as "magnetic sleep." This trance state, Martineau found from the very first sessions, effected a distinct physical improvement; from a languid haze, she awakened feeling stronger and healthier each time. She continued treatment with several mesmerists, and later with her maid, who had been trained in the procedure. Within four months, she was able to give up opiates; she began to take walks, and by December, she could go on fifteen-mile excursions. Ruddy with health, Martineau became a convert to mesmerism and a publicist of its gifts to humankind.[3] Although her enthusiastic "Letters on Mesmerism" met with derision when they appeared in the prestigious journal *Athenaeum*, still many were willing, if not eager, to undergo entrancement. When a dinner guest of George Eliot's offered to mesmerize her one evening, she readily agreed. Thackeray tried it with success. And Elizabeth Barrett, perhaps more cautious than others, confessed to Robert Browning that she believed "so much of mesmerism . . . without absolutely giving full credence to it."[4] Browning had stronger doubts, yet even he was willing to learn more. "Understand," he replied, "that I do *not* disbelieve in Mesmerism—I only object to insufficient evidence being put forward as quite irrefragable—I keep an open sense on the subject."[5] Apparently Barrett was fearful of experiencing "curious mental phenomena" while undergoing a trance, but Martineau assured her that being entranced was nothing but pleasant.[6]

Yet Barrett and others like her resisted the exposure and vulnerability inherent in the process of mesmerism. That vulnerability came only in part from one's subjection to another person's will, and even more from what one might discover by journeying into the unexplored terrain of the unconscious. The only similar experience was dreaming, but it was one thing to reflect on the meaning of one's dreams, which were experienced privately, and quite another to have a dream state induced by a mesmerist and observed by whoever else might be present. Subjects had no control over what the mesmerist might elicit from the encounter, nor did they have the ability to extricate themselves from a mesmerist trance. Mesmerism, then, felt dangerous and yet, for what it might reveal, seductive.

The potential of revelation about one's personality made "mesmeric evenings" a popular form of entertainment. These gatherings featured clairvoyant mesmerists who, by holding personal artifacts such as a letter or lock of hair, imparted something about the "moral qualities"—that is, the essential psychological qualities—of the person to whom the artifact belonged. In 1844, Margaret Fuller wrote to her friend Emerson about a gathering where the clairvoyant Anna Quincy Thaxter Parsons "read" Emerson's personality by holding one of his letters. "The writer," Parsons asserted, "is holy, true, and brave." This conclusion, which could be construed only as praise, was tempered, however, when Parsons was pressed to say more: "If he could sympathize with himself," she added, "he could with every one." Fuller was impressed: that statement, she told Emerson, "is, in my opinion, a most refined expression of the truth, whether obtained by clairvoyance or any other means."[7] Such truths, however, which might or might not concur with one's own sense of self, and which were announced to one's circle of friends or even strangers, made some potential subjects uneasy.

Despite Fuller's repeated invitations, Emerson apparently declined to attend any mesmeric evenings, although he realized fully its potential to help him understand how other people saw the world. "Each man has facts that I want, & though I talk with him, I cannot get at them for want of the clue," Emerson wrote in his journal. To enter the mind of someone far different—a "well-informed merchant," for example, or a naturalist like Louis Agassiz—would be illuminating.

When Hawthorne's fiancée, Sophia Peabody, underwent mesmerism for relief of debilitating headaches, Hawthorne bristled at the thought that someone else exerted such power over her. He saw the mesmerist's "transfusion" of spirit as a violation, a sullying of Sophia's purity and innocence. It did not impress him that Sophia claimed to be cured.[8]

Mesmerism's possible dangers form the theme of several chilling stories by Edgar Allan Poe. Although the connection of mesmerism to the supernatural or macabre fits with Poe's recurrent themes, his focus in these particular stories, published in the mid-1840s, reflects the culture's fascination with the experience, consequences, and risks of mesmerism: the relationship of the mesmerist to his subject, the possibility of mesmeric influence at a distance, the personal qualities of the "good" mesmeric subject, and the relationship of the trance state to death.

In "A Tale of the Ragged Mountains" August Bedloe is a perfect mesmeric subject: "sensitive, excitable, enthusiastic" with an imagination "singularly vigorous and creative." The vitality of his mind, however, contrasts with the debility of his body and spirits. Bedloe habitually takes large doses of morphine, apparently to alleviate the pain of facial neuralgia; morphine, no doubt, also relieved his "profound melancholy." His was a disposition, the narrator tells us, "of a phaseless and unceasing gloom." Bedloe is treated by Dr. Templeton, a disciple of Mesmer's. Bedloe and Templeton have such a strong "magnetic relation" that Templeton is able to mesmerize Bedloe even when Bedloe is unaware of Templeton's presence and even, it turns out, at a considerable distance. Besides this magnetic connection, which suggests sexual attraction on Templeton's part, there is another relationship between the two men: Bedloe, it seems, may be channeling the spirit of a man Templeton had befriended some forty years earlier, a man murdered during a riot in India. When Bedloe, wandering alone in the woods, has a vision very much like the Indian riot, Templeton is shaken. Has Bedloe become inhabited by the spirit of the dead man? Has Templeton caused Bedloe to relive his friend's experience? Has Bedloe experienced a hallucination brought on by overuse of morphine? "Let us suppose only," Templeton concludes, "that the soul of the man of to-day is upon the verge of some stupendous psychal discoveries. Let us content ourselves with this supposition."[9]

In a state of widened consciousness, the mesmeric subject might well discover transcendent insights about God, immortality, and imponderable forces. "The value of Phrenology & Mesmerism," Emerson suggested, "is not as science, but as criticism on the Church & Schools of the day; for they show what men want in religion & philosophy which has not been hitherto furnished."[10] In Poe's "Mesmeric Revelation," Mr. Vankirk, on the point of death from tuberculosis, asks a friend to mesmerize him in the hope that he can make some final discoveries. The "mesmeric exaltation," Vankirk believes, affords "profound self-cognizance" and, by extension, cognizance of levels of reality not accessible to the fully conscious individual. The story consists of a dialogue between the mesmerist and Vankirk on the nature of God ("He is not spirit, for he exists. Nor is he matter, *as you understand it*"), on the luminiferous ether (unparticled matter, Vankirk says), and on the nature of mind. According to Vankirk, "the unparticled matter, set in motion by a law, or quality, existing within itself, is thinking." If the unparticled matter is God, then living beings, and indeed all matter, are infused with God. To call God "spirit" is to utter a redundancy.[11]

Vankirk echoes Faraday's concepts of lines of force or atomic energies, concepts that Poe remarked upon in a letter to physician Thomas H. Chivers in 1844. "There is no such thing as spirituality," Poe wrote. "God is material. All things are material; yet the matter of God has all the qualities which we attribute to spirit: thus the difference is scarcely more than of words. There is a matter without particles—of no atomic composition: this is God. It permeates and impels all things, and thus *is* all things in itself. Its agitation is the thought of God, and creates."[12] This "unparticled matter," according to Poe, sometimes takes the material form of humans and other natural phenomena, transmuting back to unparticled matter after death. Matter with no atomic composition, able to permeate and vitalize, is synonymous with *electricity*.

Many readers were so eager for evidence of the powers of mesmerism that when Poe published "The Facts in the Case of M. Valdemar," the story was widely believed to be true. Reprinted many times after it first appeared in 1845, the account of a man mesmerized at the point of death "created a very great sensation," as one mesmerist wrote to Poe. "I have not the least doubt of the *possibility* of such a phenomenon," he added, "for I did actually restore to active animation a person who died from excessive drinking of ardent spirits." Elizabeth Barrett, much to Poe's delight, admitted to him that the story was so powerfully told that it had generated "dreadful doubts as to whether it can be true."[13] The story's narrator has been conducting mesmeric experiments for several months when he is invited by a friend, who is dying of tuberculosis, to investigate the effect of mesmerism on the process of death. Is the dying patient susceptible to mesmerism? How does death impede or deepen the mesmeric trance? And most puzzling, how might "the encroachments of Death... be arrested by the process?" This last question, with "the immensely important character of its consequences," intrigues the narrator most strongly, and he is eager to undertake the experiment. He is a bit surprised that Valdemar consents readily to cooperate; although Valdemar had submitted to mesmerism frequently, the narrator remarks that "he had never before given me any tokens of sympathy with what I did." Valdemar's withheld "sympathy" speaks to a significant characteristic of the mesmeric relationship: the degree to which the mesmerist and subject form an emotional and physical bond. While this bond, based on trust and even faith, was necessary for the success of entrancement, for the subject a sympathetic bond meant relinquishing power and autonomy. Unguarded, the subject readied for invasion from an invisible force.

The experiment proceeds, and as Valdemar is dying, the narrator mesmerizes him. After a short time, the attendant physicians pronounce Valdemar dead; suddenly, to everyone's astonishment, Valdemar rouses himself and speaks. "I *have been* sleeping—and now—now—*I am dead*," he proclaims. For the next seven months, the mesmerist, physicians, and nurses continue to attend to Valdemar, unsure about how to proceed with the stalled experiment. Finally, they agree to try to awaken him, realizing that if they remove him from the mesmeric state, they may also cause his final and irrevocable death. Yet try they do, and as the mesmerist makes the required passes, Valdemar emits a horrible cry: "For God's sake!—quick!—quick!—put me to sleep—or, quick!—waken me!—quick!—*I say to you that I am dead*!" Valdemar's desperate plea compels the mesmerist to make ever more rapid passes across his body until he succeeds in releasing Valdemar from his deathlike state to final death. Instantly, his body, which had remained unchanged for seven months, "shrunk, crumbled, absolutely *rotted*" upon the bed, until it was no more than "a nearly liquid mass of loathsome—of detestable putridity."[14]

The experiment on Valdemar suggested that magnetic relations were not broken by death and that in a mesmeric trance, the body might not decay. But more heartening to nineteenth-century readers was the possibility, as mesmerist Robert Collyer exclaimed to Poe, that the vital force—conveyed as animal magnetism or artificial electricity—might reanimate those who were near death or even dead. This result seemed more desirable than forestalling decay, and it became the theme of Poe's fourth magnetic tale, "Some Words with a Mummy." The mummy in question is an Egyptian count who had been dead for six thousand years before he was purchased, with other artifacts, by Captain Arthur Sabretash and given to a local museum. The narrator, summoned in the middle of the night by his friend Dr. Ponnonner, witnesses the electrification and revivification of the mummy, an opinionated fellow who proceeds to lecture his inquisitors about nineteenth-century culture and achievements compared with the glories of ancient Egypt. According to the mummy, phrenology and mesmerism had "flourished and faded" in his own time, replaced by more sophisticated science that made it possible to embalm and then reinvigorate the dead. The mummy's investigators are deflated at this news, proud as they are of their technological advances. Still, they themselves feel invigorated and "figuratively and literally" electrified by this particular experiment. The narrator attributes his own and his friends' credulousness about the mummy to "the spirit of the age, which proceeds by the rule of contraries altogether, and is now usually admitted as the solution of everything in the way of paradox and impossibility."[15]

The boundaries between the possible and the impossible were changing constantly as new discoveries, defended by clinical practice or ardent argument, made their way into the public realm. In England, John Elliotson, a physician at University College Hospital in London, became the most outspoken proponent of mesmerism, shocking some of his more conservative colleagues. Elliotson was forced to resign when he began demonstrating magnetism in the hospital's amphitheater on subjects whose honesty was suspect. Stripped of his hospital affiliation, he still continued to treat patients, among them Thomas Carlyle, Alfred Lord Tennyson, and Charles Dickens, whom he trained to be a mesmerist; Dickens exercised this talent with his family, friends, and acquaintances. His wife proved so sensitive to his powers that she could become mesmerized even when Dickens focused his magnetic emanations on someone else. In 1843, Elliotson founded a journal, *The Zoist*, devoted to "Cerebral Physiology and Mesmerism." Besides journal articles, readers had access to scores of books on the theory and practice of mesmerism, addressed to a popular audience and sometimes including instructions so that laymen could hone their own ability as mesmerists.[16]

Phrenology offered another path to self-knowledge and, when combined with mesmerism or magnetism, another therapy for physical or emotional distress. Developed late in the eighteenth century by Franz Joseph Gall, a Viennese physician, and his student Johann Spurzheim, phrenology mapped the skull, identifying more than thirty "organs" located on different parts of the brain that controlled distinct human "faculties." Phrenologists developed an esoteric vocabulary to describe these faculties, including philoprogenitiveness (love of offspring), amativeness (sexual passion), adhesiveness (friendship), and veneration (piety). There were organs, phrenologists asserted, for conscientiousness, mathematical ability, and a variety of different perceptual and cognitive abilities. By examining the bumps and shape of the skull, the phrenologist could diagnose the development of these organs, predict behavior, and explain personality. Some phrenologists magnetized or massaged these areas in an effort to alleviate symptoms or intervene in the activity of the mind, but many phrenologists aimed simply to develop some basis for understanding human psychology.

As Poe's mummy suggested, phrenology enjoyed a fairly brief spate of interest, but that interest was intense. John Elliotson became the first president of the London Phrenological Society, founded in 1824, and hosted eighteen lectures by Johann Spurzheim the following year. Throughout Great Britain, mechanics' institutes—centers of continuing education for working-class audiences—offered frequent lectures on phrenology. The phrenological bust, a human head with divisions of the skull demarcated and labeled, was an indispensable prop for such talks. Public interest was so widespread that by 1836, England had thirty phrenological societies; 64,000 copies of books about phrenology and more than 15,000 plaster phrenological heads were sold.[17]

In America, Spurzheim's visit in the early 1830s attracted enormous and respectful attention from physicians and academics. When Spurzheim died suddenly in Boston in 1832, the president of Harvard and the entire membership of the Boston Medical Association attended his funeral. He was buried among Boston's notables at Mount Auburn Cemetery. Some ten years later, when Orson Fowler, a fiery proponent of phrenology, lectured in Boston, Philadelphia, and New York, thousands filled auditoriums for courses of twenty to forty lectures.

Simon Newcomb, who later became professor of mathematics and astronomy at Johns Hopkins University, among many other prominent positions, read Fowler's *Phrenology* as an adolescent and Scottish phrenologist George Combe's *The Constitution of Man Considered in Relation to External Objects*. Combe's book, which had been published in 1828, sold an astounding 100,000 copies in Britain and 200,000 copies in America by 1860.[18] Behavior, personality, and potential, Combe argued, resulted from physiology; yet rather than presenting physiology as another form of determinism, Combe suggested that physical attributes provided the raw material upon which each person could build a plan of improvement. By understanding the extent of development of each mental faculty, individuals could exert self-direction and control by suppressing antisocial "organs" and strengthening other, more beneficial sites. In George Eliot's novella *The Lifted Veil*, for example, the narrator, when he is a child, is examined by a phrenologist and found to be overly empathetic. A scientific education, the phrenologist concludes, would strengthen weaker mental faculties. "I was hungry for human deeds and human emotions," the narrator tells us, "so I was to be plentifully crammed with the mechanical powers, the elementary bodies, and the phenomena of electricity and magnetism."[19] Science, in the popular mind, was an antidote to compassion.

Newcomb, whose upbringing was characterized by "the old Calvinistic orthodoxy in its gloomiest form," had heard too many preachers exhort parishioners about "the doctrines of the innate depravity of man, the dreadful future of the unbelievers, the futility of mere good works, and the necessity of being born again." He felt terrified, he said, by the "overwhelming force of self-condemnation." Combe liberated him from that feeling; instead of being condemned to an unalterable fate, Newcomb and other readers concluded from Combe that "all individual and social ills were due to men's disregard of the laws of Nature, which were classified as physical and moral. Obey the laws of health and we and our posterity will all reach the age of one hundred years. Obey the moral law and social evils will disappear."[20] Harriet Martineau speculated that for middle- and working-class readers, Combe's book was "the great event in the life of their minds... There they learned that their bodies were a part of the universe, made of substances and governed by laws which it is a man's duty to study and obey... Multitudes rushed into the conviction that their lot was mainly in their own power, and men set to work... to arrange their habits," especially through education.[21]

Like mesmerists, phrenologists offered both public demonstrations and private sessions. Some even offered analysis by mail, done on the basis of a daguerreotype. Walt Whitman, besides reading phrenological tracts, had his "bumps" read, taking detailed notes about his various faculties, especially that of "amativeness" and "adhesiveness."[22] Edgar Allan Poe was pleased after being examined by several phrenologists, "all of whom," he boasted to a friend, "spoke of me in a species of extravaganza which I should be ashamed to repeat."[23] When sculptor James Delville, an enthusiast of phrenology, made a life mask of William Blake, he noted Blake's well-developed "imaginative faculty." Mark Twain remembered itinerant phrenologists stopping frequently in Hannibal, Missouri, where they gave a free lecture and charged 25¢ for each examination. The phrenologists' services were hugely popular, their "formidable and outlandish" vocabulary made its way into ordinary conversation, and their "diagnoses" were unfailingly pleasing. "It was a long time ago," Twain wrote later, "and yet I think I still remember that no phrenologist ever came across a skull in our town that fell much short of the [George] Washington standard."[24] Phrenologists realized that it was good business to praise their clients, but many clients wanted more than praise. Phrenology could offer information that would help people to plan their future, even to choose a career by suggesting what kind of work matched a "natural" affinity. Some employers required that job applicants "pass" a phrenological examination that would reveal their reliability and other capacities.

Phrenology addressed a few significant questions that neither traditional medicine nor, as Emerson said, the "Church & Schools," adequately answered: What responsibility do individuals have for their illnesses? What potential do they have for self-improvement in both physical and moral health? To what extent is it possible to inquire into the unconscious? Phrenology offered a means to self-knowledge and the understanding of others: by "reading" the surface of the skull, the mind became transparent. Yet, like mesmerism, phrenology required intimacy—a sympathetic relationship—between expert and subject. As Anna Parsons noted about Emerson, sympathy was hard-won and sometimes resisted for fear of discovering something that the subject wanted to keep hidden. While sympathy implied generosity toward others, certainly a positive attribute, a sympathetic relationship also implied invasion of personal integrity.[25] Like electrical effluvia, sympathetic emanations entered one's mind invisibly, melding one's mind with another's, and subjecting it to another's will.

RIDDLES

Emanations of the mind filled the mysterious ether, according to spiritualists, and reports of supernatural manifestations filled the popular press. These reports had disturbed George Beard, and he set out to debunk them. After Beard left his practice with Alphonso Rockwell in 1874 he continued to see some patients and to write and lecture about electrotherapy. He founded a journal, *Archives of Electrology and Neurology*, to promote credibility in the field and communication among his colleagues. But he devoted most of his time and energy to studying what one colleague called "the subjective side of man": that is, the emerging field of psychology. Because psychology was not yet defined as a discipline, Beard, like other early investigators, had wide latitude in setting the boundaries of his inquiry. For Beard, that inquiry embraced "nervousness, diffidence, morbid fears, chilly feelings, blushing, and fatigue states," and also such phenomena as hypnotism, mind reading, trance states, and clairvoyance: in short, Beard decided to investigate the claims of spiritualists, and in doing so, he discovered for himself a new battleground.[26]

The last spate of enthusiasm over psychic phenomena, Beard knew, had occurred in the 1830s and 1840s, with mesmeric evenings, rappings, and testimonies of hauntings. It made sense to Beard that spiritualism had been attractive to a society of "preachers, politicians, and common people" at a time when the nation was undergoing "mental puberty, passing from childhood into youth, feeling the throbs of new desires"; the country, he said, was a "paradise of non-experts, who assumed that a perfect knowledge of a many-sided realm of thought is obtainable by an accidental glance at one side." But in the 1870s, with science a growing influence in everyday life, and experts like himself available to all, Beard was at a loss to explain why, suddenly, his colleagues and peers were "bitten, maimed, and prostrated" by a fascination with the occult. "Of all the psychological questions relating to spiritism," he said, "no one perhaps is more interesting, or has been more puzzling than this—why so many well-balanced, scholarly, generally judicious, logical, and scientific men have either enlisted under its flag or have been hangers-on of its camp."[27]

The increased attention to spiritualism that Beard noticed in the 1870s, however, had not burst into the culture but had grown gradually from a renewed interest after the Civil War. With six hundred thousand dead, grieving families searched for consolation, and the prospect of contacting spirits offered some sense of hope and comfort. Within a few years of the war's end, trance mediums boasted millions of followers.[28] Those persuaded by spiritualist claims wanted to believe that some people possessed special sensitivities to occult forces, that some people were especially sensitive to emanations from an imponderable ether, and most certainly that the spirit was not obliterated when the body died. But those who believed such phenomena, Beard knew, would reject explanations of mind and behavior that were based on cells, nerves, and organic processes; they would reject psychology as a science. Spiritualists would undermine his own authority as an expert on the mind.

By the late 1870s, Beard had moved away from the position he took when reviewing Noah Porter's *The Human Intellect* in 1869. Then he had allowed for the existence of an immortal soul, "identical with the vital force, or . . . correlated to it and to all the other forces of the body."[29] Now he seemed to share the same "crude materialism" that he had condemned in other scientists. Beard surely was alert to the power of mind over matter, and like other electrotherapists, he knew that if his batteries failed, patients often testified that the treatment was as effective as when they actually received an infusion of current. When he and Edison tested the physiological effect of the etheric force, Edison had claimed that when the force entered his body, he felt a contraction of muscles in his tongue and neck, causing his head to nod; but when Beard surreptitiously broke the connection, Edison experienced the same responses. "It was a case of mind acting on body," Beard concluded, "he expected some effect and unconsciously produced it himself." One night during the experiments, Edison and two assistants felt sick and attributed their illness to the etheric force, but again, Beard diagnosed their feelings as psychosomatic.[30] Beard had little patience for those claims, especially when they involved such phenomena as hypnotism, mind reading, and communication with spirits. "If trance, the involuntary life, and human testimony, were understood universally as they are now beginning to be understood by students of the nervous system," he wrote, "there would not, could not be a spiritist on our planet; for all would know that spirits only dwell in cerebral cells—that not our houses but our brains are haunted."[31]

To debunk clairvoyance, he conducted an experiment with some of his patients, asking them not to tell him anything about their symptoms but to let him guess, "after the manner of a clairvoyant," what ailed them. To his patients' amazement, he was always right, "and with good reason," he said, "for nothing that I could do prevented them from telling me, although I asked them no questions; unintentionally and unconsciously, they would guide me at every stage of the interview." He was certain that clairvoyants made their own diagnoses in the same way: by listening for hints about a person's life and by honing their sensitivity to subtle gestures and glances. Of course, he was aware that clairvoyants were not above resorting to trickery: some used accomplices to spy on clients in a waiting room or listen to their conversations beyond closed doors; and others offered nothing more than generalizations that could apply to anyone. Based on his experiment, subjective though it was, and his unflappable convictions, Beard was convinced that psychic force could not exist. "Science," Beard announced, "thus becomes the real and only clairvoyant; only through the eyes of science is it given to man to read the future."[32]

Beard waged a vociferous attack against the Russian spiritualist Helena Blavatsky, who had come to New York from Paris in 1873, trailing scandal behind her. In 1874, she heard about some occult performances in rural Vermont and was enticed to see them for herself. William Eddy—a poor, uneducated, and, visitors discovered, unsociable farmer—had a talent for spirit materialization. Eddy held séances in a second-floor hall in his farmhouse where he would sit in a closet darkened by a blanket hung across the door and wait for spirits to appear. It took no time at all before a figure of a dead person stepped from behind the blanket, appearing for a moment like "an animate statue," as one visitor described the vision, before it dissolved into invisibility. The identity of the figures changed depending on the spectators; when Blavatsky visited, dead Russians appeared, as did one of her deceased uncles. Because Eddy was so unprepossessing, no one imagined that he could be duplicitous; his ability to materialize spirits seemed to be the real thing.[33]

Beard, though, had strong doubts. Incensed by reports of the Eddy séances, he went to Chittenden to see them for himself and published his critique as a belligerent letter in the *New York Daily Graphic*, where Blavatsky's testimony had appeared. Beard claimed that with a few dollars worth of drapery, he himself could produce the so-called apparitions. Blavatsky's angry reply earned her the attention she craved, but Beard's argument carried considerable weight. He went on to lecture about spirit materialization, mesmerism, and a topic that was becoming his favorite: mind reading.

Mind reading, he claimed, was nothing more than the "reader's" sensitivity to almost imperceptible muscle movements. The phenomenon should be called muscle reading, he said, and took up the battle against Jacob Brown, a mind reader so famous that he captured the attention of the Yale faculty, who invited him to demonstrate his abilities in a public test. Beard attended, certain that the mind reader was a fraud, and he found himself confronting the credulous faculty, who simply wanted to believe.

While Beard argued that trances and clairvoyance could be explained physiologically and psychologically, Edison investigated a few spiritualist precepts in his laboratory. Edison insisted that he could not "conceive of such a thing as a spirit" with "no weight, no material form, no mass . . . I cannot be party to the belief that spirits exist and can be seen under certain circumstances . . . The whole thing is so absurd." Therefore, if the personality did exist after the body's death, it must have some material form, however slight, which a "super-delicate" instrument might detect. "For my part," he admitted, "I am inclined to believe that our personality hereafter will be able to affect matter." He was certain that spirits would not communicate with the living through means as "mysterious, or weird" as a human medium, causing table tilting or rapping, but through an extremely sensitive scientific apparatus. "If what we call personality exists after death," he said, "and that personality is anxious to communicate with those of us who are still in the flesh on this earth, there are two or three kinds of apparatus which should make communication very easy." He said that he tried to construct such a device, "in the nature of a valve," he explained, "so delicate as to be affected, or moved, or manipulated . . . by our personality as it survives in the next life."[34]

Edison's speculation about the possibility of "infinitesimally small" units of life made him more sympathetic to Blavatsky than Beard was. When Beard was blasting her in the pages of the *Graphic*, Edison received a dinner invitation from her to talk about his etheric force. He declined, but he did tell one of Blavatsky's associates that he was involved in an experiment of his own, testing whether the will had physical powers. He suspended a pendulum on the wall of his workshop, attached various wires from the pendulum to his forehead, and tried to move the pendulum "by will-force." The experiment apparently failed, or surely it would have merited a patent. Edison's suspicions about spiritualism did not prevent him from filling out an application of membership in Blavatsky's Theosophical Society, nor from donating a phonograph to be sent to the Bombay branch.[35]

Edison's flirtation with spirit communication and psychic force and Beard's fierce opposition to spiritism stem from the same source: their recognition that the imponderable ether, with all of its supernatural connotations, prevailed as a vivid, potent, and awe-inspiring concept in the popular imagination. Beard was right in thinking that the brains of nineteenth-century men and women were haunted: even educated people like himself cherished the idea that something in the world always would be impenetrable and magical but nonetheless real; many resented scientists' insistent deriding of the occult. They hoped that science, instead, would change to look beyond the empirical and accept the existence of what William James called "wild facts." "Any one will renovate his science," James advised, "who will steadily look after the irregular phenomena."[36]

Two years after his initial battle against spiritualism, Beard had his convictions shaken, if not broken down. Although his essays and lectures led readers to characterize him as "an earnest hater" of spiritualists' claims, he continued to investigate clairvoyants, suggesting that the "riddle of the Universe" remained unsolved for him. Surprising, and perhaps gratifying, results came from an experiment with a young woman who claimed the power to read any printed material that was put in contact with her forehead. With her eyes bandaged by Beard, with Beard pressing a piece of paper upon her forehead, the young woman proceeded to fulfill her promise. The demonstration, reported respectfully and at length in *The New York Times*, convinced Beard of the mind's apparently supernormal power. Sight and touch were not involved; facial nerves could not perceive letters written in pen and ink. The only conclusion to be made was that the mind "can act outside and apart from the body, and the materialistic theory that the mind is identical with the gray matter of the brain is utterly overthrown."[37] If most spiritualism was trickery and fraud, this experiment suggested that science needed to be renovated, expanded, and energized.

A WORLD UNSEEN

Beard was not alone in attempting to bring scientific objectivity to an investigation of the occult. At the same time that Edison illuminated a quarter mile of Manhattan, a group of prominent physicists and philosophers in Cambridge, England, gathered in search of a different kind of illumination: into the invisible realm of the ether and the mind. The London Society for Psychical Research (SPR) was an outgrowth of the Cambridge Ghost Society, founded in the 1850s by several noted clergymen, one of whom, Edward White Benson, later became archbishop of Canterbury. Revived and renamed, the new organization of psychical researchers was mindful of ghosts and spirits, but also of such phenomena as clairvoyance and telepathy (a term one of its members coined). Like Beard, the members of the SPR were impatient with anecdotal testimony about such phenomena, and they believed that by applying scientific rigor to their investigations, they could address questions about the existence of the soul, human immortality, and the ability of humans—alive or dead—to communicate at a distance or influence another person through thought alone. Less ambivalent than Beard about their goals, they hoped to collect enough evidence to make a case for the existence of an unseen universe containing powerful, mysterious, ethereal energies.

The founding members of the SPR included eminent British intellectuals: physicists William Barrett, Sir Joseph John Thomson, and Sir Oliver Lodge; philosopher Frederic Myers and classicist Edmund Gurney; Arthur J. Balfour, who later became prime minister, and Balfour's brother-in-law, the mathematician and philosopher Henry Sidgwick. Two years after the London society was inaugurated, William James, Simon Newcomb, and others inaugurated the American branch, whose membership boasted clergymen, professors, physicians, and college presidents: Reverend Phillips Brooks; publisher Henry Holt; professors G. Stanley Hall (Johns Hopkins) and Asa Gray and Josiah Royce (Harvard); and Andrew White (president of Cornell). Theodore Roosevelt was an associate member; Alexander Graham Bell contributed to the group's investigations. Thomas Edison collected *The Proceedings of the Society for Psychical Research, 1882–83* in his library.[38]

The SPR organized itself into committees, whose mission it was to amass testimony about a specific category of supernormal phenomena. The Reichenbach committee, named for the discoverer of the odyl, investigated the effect of magnets on the human body. The committee on clairvoyance attended countless séances; the committee on trance states underwent, administered, and observed hypnotism; the committee on hallucinations compiled a census of twenty-five thousand testimonies. Based upon these copious interviews and observations, the committee members were charged with formulating conclusions about the cause of the phenomena. They were especially interested in telepathy, implicated as it was with imponderable energies and the power of the will to transmit thought and to influence others.[39] Successful telepathic experiments, proving the strength of human will, raised the question of how thoughts traveled from one mind to another by a "hypothetical force or mode of action" that Myers—the society's lexicographer—called, vaguely, *telergy*. Discoveries by Helmholtz and others revealed that chemical processes in the brain and nerves generated electricity, leading psychical researchers to suppose that telergy and willpower were electrical. Subjects able to send their thoughts strongly and subjects able to concentrate intently on receiving those thoughts were endowed with particularly energetic wills.

But subscribing to the electrical nature of will, and more generally of consciousness, did not lead psychical researchers to hold a purely materialist belief in the nature of the mind. William James, for example, wholeheartedly rejected the idea that consciousness was produced solely by currents in the brain and nerves, governed by mechanical laws. The range of conscious states, the ability of someone to pay attention to one thing and not to another, suggested to James that "a belief in free will and purely spiritual causation is still open to us."[40] As he saw it, consciousness was best explained by both a production theory and a transmission theory. Surely the brain produced certain thoughts and feelings in reaction to stimuli; but other thoughts, images, and feelings existed "ready-made in the transcendental world" and could be perceived by anyone receptive to them. "The word 'influx,' used in Swedenborgian circles," he said, "well describes this impression of new insight or new willingness, sweeping over us like a tide... We need only suppose the continuity of our consciousness with a mother sea, to allow for exceptional waves occasionally pouring over the dam."[41] The mother sea, the mysterious ether, contained vital energy.

Still, if connection to ethereal forces explained some manifestations of consciousness—especially the talents of mediums, which were James's area of investigation—such forces seemed not to account fully for hypnotic states and automatism. With these phenomena, researchers discovered a notable absence of willpower, similar to what neurasthenic patients complained about to their physicians. Subjects in a trance state could be controlled in both action and speech by their hypnotist; and in experiments in automatic writing, the subject produced words and images that he or she did not intend or will to write.

Automatism, according to Myers, referred to "such images as arise, as well as such movements as are made, without the initiation, and generally without the concurrence, of conscious thought and will." Sensory automatism involved visual and auditory hallucinations; motor automatism involved words written or spoken without intention. One way that the SPR investigated automatism was with a device called a planchette, a plank of wood upon which the subject rested his or her arm while holding a pencil. Distracted by a conversation with the investigator, the subject would relax and the suspended arm would eventually begin to write; the communications, sometimes incoherent and sometimes in strange signs or language, were not the product of intention or will. Where, then, did they come from?

Philosopher Frederic Myers had a tantalizing theory: "I ascribe these processes," he said, "to the action of submerged or subliminal elements in the man's being." By subliminal, Myers explained, he meant neither "reflex cerebral action" nor "unconscious cerebration": reflex implied cellular or chemical activity; cerebration, rational thought. What Myers suggested was far different: every individual consisted of a layered self, with some layers hidden beneath the threshold of consciousness, but constituting, nevertheless, the individual's identity. Subliminal "excitations," he argued, were kept submerged, not because they were weak or inconsequential but because of "the constitution of a man's personality. The threshold," he added, "must be regarded as a level above which waves may rise,—like a slab washed by the sea,—rather than as an entrance into a chamber."[42]

The SPR's investigations focused on energies—transmitted, repressed, subliminal—and though they inquired into what James called "irregular phenomena," they reflected on the kinds of experiences that anyone might have: transient feelings, premonitions, overwhelming waves of emotion, unexpected intrusions of memory into the present moment, dreams, and nightmares. Their findings were disturbing and contradictory: it might be possible to control others through a force of one's will, but not necessarily to control one's layered self; the will might be a product of the brain and nerves, yet could be assaulted by, and even connected to, ethereal energies; scientific method could yield insight into the spiritual realm, and as it did, it could undermine the essential tenets of science. Psychical researchers created a world in which individuals were clinging precariously to a safe harbor of rationality, able to exert only limited self-control, vulnerable as they were to waves from the mother sea, filled with imponderable energy, and to inner surgings from the depths of the unconscious.

Investigations into trance states, James said, "have broken down for my own mind the limits of the admitted order of nature. Science, so far as science denies such exceptional occurrences, lies prostrate in the dust for me; and the most urgent intellectual need which I feel at present is that science be built up again in a form in which such things may have a positive place."[43] James's critique of science as merely another form of faith, a critique echoed by others, had implications beyond psychical research: the distrust that Edward Van Deusen had identified as the most significant symptom of neurasthenia surfaced in the public's response to scientists' claims about the body, medicine, and the electrical force.

CHAPTER 8

The Inscrutable Something

British medium John Beattie (left) leads a spirit circle in which psychic force emanates from the sitters' hands. Spirit photographs, such as this carte de visite (Bristol, England, 1872), were promoted as material proof of occult energies.

Eugene Rochas Papers, American Philosophical Society

Materialism means simply the denial that the moral order is eternal, and the cutting off of ultimate hopes.

William James, *Pragmatism*

In 1834, the British journal *Quarterly Review* remarked upon "the want of any name by which we can designate the students of the knowledge of the material world collectively." The British Association for the Advancement of Science (BAAS) took up the challenge each summer for three consecutive years, but its members could not agree. "*Philosophers* was felt to be too wide and too lofty," especially by Samuel Taylor Coleridge, who, at one of the meetings, objected adamantly to the term; similarly, the BAAS thought that "*savants* was rather assuming." Finally, one member proposed that "by analogy with *artist*, they might form *scientist*, and added that there could be no scruple in making free with this termination when we have such words as *sciolist*, *economist*, and *atheist*—but this was not generally palatable." Six years later, polymath William Whewell ended the debate. A minister of the Church of England, master of Trinity College at Cambridge, member of twenty-five scientific societies in England and on the Continent, philologist, and champion of induction as a valid method of knowing, Whewell already had contributed such terms as Eocene, Pliocene, and Miocene to the field of geology. "We need very much a name to describe a cultivator of science in general," he proclaimed. "I should incline to call him a Scientist."[1] The term, once and for all, entered the English language.

As *scientist* came to replace *naturalist*, the difference proved troubling. In 1833, when Emerson first visited the Cabinet of Natural History in the Jardin des Plantes in Paris, he came away filled with wonder at "the inexhaustible riches of nature. The Universe," he wrote in his journal, "is a more amazing puzzle than ever as you glance along this bewildering series of animated forms." And equally amazing was the undeniable connection he felt to these strange and fabulous beings. "Not a form so grotesque, so savage, nor so beautiful but is an expression of some property inherent in man the observer—an occult relation between the very scorpions and man. I feel the centipede in me—cayman, carp, eagle, & fox. I am moved my strange sympathies, I say continually, 'I will be a naturalist.'"[2] As a naturalist, Emerson would join others in search of a theory of nature that could explain the "occult relation" among all natural phenomena: in short, a theory that revealed God as the ultimate creative force, professed sympathy between humans and all other living entities, and enriched one's sense of humanity.

If, along the path to a unified theory, the naturalist made discoveries—of new flora and fauna, of physical or chemical behavior—humankind might enjoy some practical benefits. But the naturalist's true vocation was not invention or innovation, but affirmation of a divine intelligence. That vocation, from the moment that Whewell proposed the term *scientist*, seemed increasingly eroded as researchers focused on a partial view of nature according to their special vantage. A botanist or a physiologist, a neurologist or a chemist, delivered one theory and another into the world, fragmenting rather than unifying nature. Those discoveries made for sensational news stories, disseminated by the popular press, sometimes accurately, often not. "The immense amount of valuable knowledge now afloat in society enriches the newspapers," Emerson wrote sardonically in 1848, "so that one cannot snatch an old newspaper to wrap his shoes in, without his eye being caught by some paragraph of precious science out of London or Paris which he hesitates to lose forever. My wife grows nervous when I give her waste paper lest she is burning holy writ, & wishes to read it before she puts it under her pies."[3] Emerson's wife was not the only one growing nervous about the onslaught of scientific information. With more than two hundred professional associations founded between 1870 and 1900, it seemed that there was an expert opinion on everything, opinions that often contradicted one another, supported by evidence that the public felt inadequate to evaluate. Especially in the sciences, expertise seemed increasingly to convince nonexperts that investigating natural phenomena was an esoteric activity and that theories could not be evaluated by common sense. Most certainly they convinced a wary public that scientists were intent on reducing nature to mere matter.

No one resisted a materialist universe more eloquently than William James. In talks that attracted hundreds of listeners and in articles in popular magazines such as *The Atlantic Monthly* and *Scribner's Monthly*, James responded to his contemporaries' concerns about the consequences of scientific research; the limits of scientific authority; the possibility of affirming religious faith in a world that privileged the empirical; about, as he put it, "the contradiction between the phenomena of nature and the craving of the heart to believe that behind nature there is a spirit whose expression nature is."[4] Trained in chemistry at the Lawrence Scientific School in Cambridge and graduated from Harvard's medical school, James taught anatomy as his first academic position, established a laboratory for psychological research, and in 1890 published a definitive textbook on psychology. These experiences, as well as his engagement in psychical research, offered him ample opportunity to reflect on the enterprise and assumptions of science, and their connection to his own longing for spiritual faith.

For James, science itself was grounded in faith: that one method of verification, involving repeated testing and gathering of large amounts of empirical data, would lead inevitably to a discovery of truth. The scientific investigator, he said, "has fallen so deeply in love with the method" that "truth as technically verified" has become the only goal of science.[5] James was skeptical, though, about science's claim of objectivity. Science should imply, he said, a "certain dispassionate method. To suppose that it means a certain set of results that one should pin one's faith upon and hug forever is sadly to mistake its genius, and degrades the scientific body to the status of a sect."[6] As James saw it, the sect held a "certain fixed belief,—the belief that the hidden order of nature is mechanical exclusively, that non-mechanical categories are irrational ways of conceiving and explaining such things as human life." Other ways of thinking—religious, ethical, poetical, emotional—were based on "the personal view of life" rather than the mechanical, the "romantic view" rather than the rationalistic. For the scientist, the "chronic belief of mankind, that events may happen for the sake of their personal significance, is an abomination; and the notions of our grandfathers about oracles and omens, divinations and apparitions, miraculous changes of hearts and wonders worked by inspired persons, answers to prayer and providential leadings, are a fabric absolutely baseless, a mass of sheer *un*truth."

Nature, though, provided evidence of many such experiences: experiences that were "capricious, discontinuous, and not easily controlled"; experiences that depended upon "peculiar persons for their production." These "exceptional occurrences," James said, required that science "be built up again in a form in which such things may have a positive place."[7] But James knew well that most scientists resisted inquiry into whatever transcended matter. If telepathy could be proven, a biologist once told James, "scientists ought to band together to keep it suppressed and concealed. It would undo the uniformity of Nature and all sorts of other things without which scientists cannot carry on their pursuits."[8] James thought otherwise. Humanity needed to believe that what existed had meaning, that "the so-called order of nature, which constitutes this world's experience, is only one portion of the total universe," that beyond the empirical there existed "one harmonious spiritual intent."[9] That is, humanity needed to believe in the existence of God.

"A world without God in it to say the last word, may indeed burn up or freeze," James admitted, alluding to recent theories about the fate of the planet, "but we then think of him as still mindful of the old ideals and sure to bring them elsewhere to fruition." Tragedy and despair, then, are only transitory, and "not the absolute final things... Materialism means simply the denial that the moral order is eternal, and the cutting off of ultimate hopes; spiritualism means the affirmation of an eternal moral order and the letting loose of hope."[10] A world of promise and hope, according to James, is a world that energizes, that justifies action, that urges the exercise of free will, and, simply, that promotes a feeling of joy. Only belief in an Absolute, God, Spirit, or Intelligent Design can ensure such a world.

"Science says things are," James noted, "morality says some things are better than other things; and religion says essentially two things. First, she says that the best things are the more eternal things, the overlapping things, the things in the universe that throw the last stone, so to speak, and say the final word... The second affirmation of religion is that we are better off even now if we believe her first affirmation to be true."[11] This desire to invest the empirical with the spiritual reflected a fear that science was essentially amoral, incapable of offering moral guidance; this fear created tension and unease about many new discoveries and inventions. Sometimes that tension was relieved by portraying a scientist or researcher as a magician, as Edison was early in his career, or as an artist, or even as a special variety of priest.

HALOS OF LIGHT

Nearly a decade after the press anointed Edison the Wizard of Menlo Park, reporters contrived the same image for another young electrician and inventor, thirty-two-year-old Nikola Tesla. As Edison's reputation became tarnished by lawsuits and rumors of greed, Tesla—young, eager, brilliant, with what reporters called a "poetic temperament"—vied for his place as magus. Born in Croatia, Tesla was trained at Graz Polytechnic Institute and Prague University. Although his parents wanted him to become a priest, Tesla, wrote an admiring contemporary, "felt himself destined to serve at other altars than those of his ancient faith, with other means of approach to the invisible and unknown."[12] To pursue his passion, Tesla emigrated first to Paris, where he worked for Edison's French affiliate, and then, in 1884, to America. For less than a year, he worked for Edison, who at first was impressed with Tesla's imagination and energy. But the relationship soured, not least because they differed about the relative merits of direct and alternating current. Edison favored direct current, where a generator causes electricity to flow in one direction to whatever the generator powers. The advantage of direct current, according to Edison, was its continuous production of fairly low voltage, 110 volts. Tesla favored alternating current, powered by a special generator—an alternator—that causes electricity to flow back and forth from high to low voltage; in alternating current systems, high voltage, say, of 1,000 volts, is reduced by transformers before it reaches whatever it powers—lights, for example, or machinery. Alternating current could provide an efficient electrical system in which a power station served a larger radius than a direct current system; power could be generated more cheaply; and with good insulation and transformers, the energy would be as safe for consumers as direct current. Edison was not persuaded by Tesla's argument, and not necessarily on scientific grounds. Because he had devoted himself to direct current technology, because he stood to profit by selling generators as part of his electrical systems, he was not inclined to support a system that required fewer power stations.

Besides these professional differences, Edison also was disconcerted to find that Tesla needed less sleep than he did—two hours at most, with no catnaps on the laboratory table. Although Tesla's talents were undeniable, to Edison he seemed erratic and disturbingly exotic. Finally Tesla quit, accusing Edison of cheating him out of a considerable sum of money. Soon after, he was hired by George Westinghouse, who recognized the financial advantage of alternating current.

Because low voltage seemed intuitively safer than high voltage, Tesla took it upon himself to reassure the public that they had nothing to fear. In his demonstrations, he showed the safety and practical application of alternating currents by allowing an extraordinarily high voltage, vibrating at a million times per second, to pass through his body, producing "dazzling streams of light." After the demonstrations, his audiences were amazed to see that Tesla's body and clothing continued, ethereally, to emit "fine glimmers or halos of splintered light."[13] Tesla's lecture at London's Royal Institution was particularly astounding, and his eminent audience felt certain that Tesla's work would propel them into the electrical future; here, they thought, was "a scientific explorer who, if health and life be granted him, will travel fast and far." If the principles of Tesla's discovery were put into use, the system of incandescent lighting that hardly had made its way into homes already would become obsolete. Tesla's conception of lighting did away with filaments and wires, sparks and shocks. Instead, Tesla envisioned in every room "an electric field in a continual state of rapidly alternating stress . . . while vacuous bulbs or phosphorescent globes and tubes, without care or attention, would shed a soft, diffused light, of colour and intensity arranged to suit the most luxurious fancy." Tesla was certain that this alternating stress would cause "no unpleasant effects" on the body or the mind, but even a laudatory report in *Nature* betrayed some skepticism. "It would be interesting also to know," the reporter mused, "whether, after all, habitual dwelling in a region of electric stress rapidly changed from one extreme of high intensity to the opposite, produced very slow physiological effects which could be traced in the improved health and longevity of the persons so dwelling, or the reverse."[14] Despite this gap in knowledge, Tesla's discoveries generated a certain amount of acclaim and even awe. For *Century*'s associate editor Robert Underwood Johnson, ordinarily a temperate writer, Tesla's laboratory inspired breathlessly romantic verse: there, Johnson extolled, one could sense "blessed spirits waiting to be born." Tesla would "unlock the fettering chains of Things" to lead humankind to "The Better Time; the Universal Good."[15]

Other writers, too, perpetuated this image of Tesla as the latest liberator of knowledge and bringer of gifts for humankind. As one journal noted euphorically, Tesla "has almost within his grasp the key that will force from reluctant nature the disclosure of one of her greatest mysteries—the production of light without heat."[16]

GODS OF ELECTRICITY

Articles about Tesla and Edison made the public aware of electricians' disagreement over the kind of current that would be safe and efficient for public use. That disagreement, and articles about whether or not alternating currents caused stress to the body, fueled anxiety about electrification and a deeper concern as well: the wisdom of forcing secrets from nature. "An Electrical Study" is, superficially, a typical nineteenth-century love story, in which a woman, aspiring to free herself from the constrictions of her roles as wife and mother, submits to the love of a good man and decides that she will be happiest at home. Electricity, however, gives this familiar plot a new twist. Girard Channing is an electrician working in the Patent Office, where he meets the attractive young Enid Wentworth, who applies for a job, hoping to help out her brother, an inventor.[17] Because Channing is a scientist, he takes hardly any interest in Enid: science is both his love and his religion; he worships the god of Electricity. This wondrous power, the narrator tells us, "is typified by Mercury, the swift-footed messenger of the gods, obedient, serviceable, running to the ends of the earth on our errands with inconceivable rapidity. And in another view it is like Diana, beautiful, but dangerous and untamed, who will not be trifled with, and whose touch is death to the Actaeon that comes too near." Channing is supercilious toward Enid, herself characterized as overly sensitive and, like some of her contemporaries, too eager to enter the workforce. Part of her motivation is, of course, to help her brother, but she also feels "vague longings to be of some use in the world," while knowing that she has no skills to make her useful. She knows, also, that by entering the workforce she risks losing the qualities of passivity and receptivity that make a woman attractive to men. "A man's looks, such as they are," Channing tells her, "are not affected by his business; but a few years of routine work is death to a woman's beauty." Enid wonders why, if women's work must be limited to being wife, housekeeper, and mother, some women fulfill those roles so badly, but Channing blames the "breaking down rather than fulfillment of their natural vocation" on "modern civilization, with its hard requirements and comparatively few helps." With this retort, Enid is reduced to silence, forced to compare what she deems is her inferior intellect to Channing's "solid qualities." "I amount to so little as I am," she tells herself, "yet there is something in me worth cultivating; and he is the man who could bring it out if he only cared to try." The tension between prospective lovers, familiar to readers of magazine fiction, is complicated here by Channing's identity as a scientist: cold, rational, unsentimental, obsessed with probing and dominating nature.

Yet Channing's emotions are stirred when he hears Enid playing the piano. Her performance leaves him "spell-bound," and he acknowledges Enid's inherent artistic power, incompatible, he believes, with intellectual power. "Why will you pose as a strong-minded woman and a worker?" he asks her, when she might better nurture her artistry and sensitivity. That sensitivity extends to her appreciation of nature, far different from Channing's. Nature, according to Enid, is shy, modest, and mysterious, an inspiration to the poetic imagination; but for a scientist, Nature is a wrathful temptress, daring men to conquer and subdue her. When Channing finally confesses his love to Enid, she retorts, "You love science more, Mr. Channing... Science! she is so deep,—not seen through in a minute, like us shallow girls, but always offering something new." Although Channing protests that he has room in his heart to love science and to feel "another emotion," Enid agrees: "Yes, you dearly love power, which is perhaps more human, if no more tender." He has tried to control her mind, she tells him, and what he calls love, she sees as fulfillment of his "masculine vanity" by exerting control over her heart.

With this standoff, the two part, Enid to return to her brother, Channing to attend to an electrical problem in a nearby town. There, he is electrocuted: the vengeful Diana destroys her arrogant suitor. Once he is dead, of course, Enid realizes that he was the love of her life, capable of inspiring her passion, although he had "erased all sentiment from his dictionary." Channing's love for electricity, it seems, could have energized Enid as much as if he administered electrotherapy. "Some possibilities in her died with his death," she reflects. "Her nature, true in its theories, strong in good impulses, but unequal to continued strain, needed the tonic of his stronger, more concentrated character, and in time would have profited by it, and he by the more spiritual qualities of hers." Eventually, weighing her options for the future—a hostile workplace where she risks losing her beauty—she marries another electrician, older and less fiery than Channing, a man "of ripe judgment and cultivation," who understands that she will grieve her loss forever, and forever feel "a mingled attraction and horror" for electricity.

The horror was generated partly by electricity's power to kill, partly by its power to dehumanize. Although Channing did claim to feel tenderness toward Enid, she believed that science, power, and control over nature—even her own human nature—were his deeper passions. In Enid's eyes, Channing had been dehumanized by science, and, if he had his way, she would be another victim. This fear of dehumanization, of electricity's power to turn humans into automatons, recurs in a novel as macabre as *Frankenstein*, Auguste de Villiers de L'Isle-Adam's *Tomorrow's Eve*, first published in serial form in 1885–86.[18]

Villiers had admired Edgar Allan Poe since the 1850s, when he first read Baudelaire's French translations, and his own tales of terror and the fantastic earned him praise as Poe's French heir. But Villiers's fiction was informed also by his reading of Hegel and his fascination with spiritualism, mesmerism, and the occult. Frequently, his characters are neurasthenic men with acute sensitivity to a reality beyond the empirical and an unshakable conviction that the soul must transcend death. Predictably, Villiers, like many of his readers, was suspicious of the materialist claims of science and technology. *Tomorrow's Eve* features Thomas Edison as the creator of the perfect mechanical woman, in every way but one the image of the most desirable woman on earth: his Eve lacks a soul. Unlike Victor Frankenstein, who was so daunted by female physiology that he could not create a mate for his Creature, Edison draws upon "electromagnetic power and Radiant Matter" to produce an automaton so lifelike that, he claims, she could deceive her own mother.

The protagonist, Lord Ewald, in love with the foolish and vain Alicia Clary, sees clearly the drawbacks of a relationship with her. Thomas Edison, claiming godlike powers, offers to help, promising to produce Miss Clary "not simply transfigured, not just made the most enchanting of companions, not merely lifted to the most sublime level of spirituality, but actually endowed with a sort of immortality."[19] Edison convinces Ewald that love is narcissistic: he is in love with the qualities he desires in Alicia, and not the qualities that she herself possesses. By transplanting Ewald's desires into a new being, Edison will create, in short, Ewald's ideal woman. Edison assures Ewald that the woman he will fashion bears no relationship to previous automatons; new techniques of what he calls "identification" render perfectly "potent phantoms, mysterious presences *of a mixed nature*... the grace of her gesture, the fullness of her body, the fragrance of her flesh, the resonance of her voice, the turn of her waist, the light of her eyes,... the individuality of her glance, all her traits and characteristics, down to the shadow she casts on the ground."[20] But instead of the crassness of Miss Clary's personality, Edison will substitute refinement and sensitivity; the replacement of her soul, he says, will render her "capable of impressions a thousand times more lovely, more lofty, more noble."[21]

Ewald waits in eager anticipation, while Edison closets himself in his laboratory. Creating a woman takes time, and during Edison's apparent disappearance, rumors fly that he is ill, or is at work on a groundbreaking discovery, or has died. A gas company sends spies to ferret out the truth. Finally, Edison summons Ewald to meet the new and improved Miss Clary. She is, Ewald discovers, indeed a lovely creature. Besides her fine sensibilities, she has the added attraction of not ever needing to eat. Lozenges and pills suffice; and Ewald can set her in motion by pressing rings on her fingers. Her pearl necklace, too, controls various functions. Surely she is a masterful creation, but Ewald is disturbed: "She isn't a *being*!" he exclaims.

"As to that," Edison reminds him, "the world's most powerful minds have always been asking themselves what is this notion of Being, considered in itself. Hegel... has demonstrated that when you consider the pure idea of Being, the difference between that and Nothing is simply a matter of opinion."[22]

Late in the century, the world's most powerful minds seemed to be inventors and engineers, focused on technology rather than moral philosophy. The fictional Edison's chilling comment evoked the same fear that Emerson had expressed nearly forty years earlier, when electrical technology was still in its infancy, and he worried about the manipulation of animal magnetism through mesmerism and phrenology. "There must be a relation," he wrote in his journals, "between power and probity. We have, no doubt, as much power as we can be trusted with."[23]

THE DOCTOR IN THE WITNESS BOX

Contentions among scientists to explain reality became intensely disturbing when focused on the mind and the persistently puzzling question of what it means to be human. A fragmented group of experts—physicians, neurologists, psychologists, lay electrotherapists, mind curers, chemists—offered responses to that question that were confusing and contradictory, based on a wide range of evidence, or sometimes no evidence at all. To what extent was human behavior determined by heredity? To what extent did humans have free will? Could autopsies of diseased brains or dissections of animals' brains yield insights about the human mind? If consciousness was not tangible, how could it be investigated? George Beard had responded to such questions throughout his career, and in the summer of 1881, he became involved in a curious and much publicized criminal case that brought his ideas into public debate, a debate that revolved as much around the issue of scientific expertise as it did on the workings of the mind.

On Saturday morning, July 2, 1881, James A. Garfield, who had been inaugurated in March as America's twentieth president, arrived at the Baltimore and Potomac station in Washington, D.C., to take a train north to Williamstown, Massachusetts.[24] He planned to attend commencement at Williams College and then, with his family, leave for a vacation. But two shots rang out in the quiet station, one piercing his back, the other grazing his arm. Garfield crumpled to the platform, blood seeping through the fabric of his suit. He was surrounded immediately, moved onto a mattress, and given a sip of brandy. Several physicians arrived quickly and inspected the wounds; yet almost an hour passed before Garfield was transferred by police ambulance back to the White House. Physicians visited for the next few hours, agreeing to wait until Garfield's condition improved before probing for the bullet. Meanwhile, they administered morphine and kept an uneasy watch.

Back at the station, police officer Patrick Kearney quickly had taken the shooter into custody. Grabbed by a ticket taker and depot watchman, Charles Guiteau admitted, "I did it. I will go to jail for it." In his pocket were several letters in which he also attested to the deed: one was addressed to the White House, another to General Sherman of the War Department. Once he was in the police station, Guiteau, slight, unkempt, and decidedly agitated, became unusually talkative. He told the police that he was a lawyer, a member of the Stalwarts, a political faction opposed to Garfield, and that he had shot the President in order to unite the Republican party and achieve for himself some political gain.

In the next days and weeks, as Garfield at first rallied, then worsened, and on September 19 died, the country struggled to understand what had occurred and why. Garfield's assassination inspired grief and uneasiness, a loss of faith in the steadiness of the world, a sense that the unexpected could, and did, shatter one's certainty and security. It seemed inexplicable that so mild and innocuous a president as Garfield would be vulnerable to assassination, and as discussion of the event played itself out in the press, the courtroom, and even the pulpit, two issues emerged as the focus of popular concern: the question of Guiteau's sanity—and, therefore, the boundary between sanity and insanity—and the existence of the will.

Immediately the stunned nation asked why Guiteau had shot Garfield. In a democratic republic, there were ample measures for anyone, Guiteau included, to have his views heard and concerns addressed; even if he were filled with hatred toward Garfield, he had only to rise up and declare himself. Why, then, did he resort to murder? Reports from the jailhouse where Guiteau was being held portrayed the assassin as a ne'er-do-well, a loafer, an abusive husband, a philanderer, an eccentric: But was Guiteau insane? Who would be able to make the diagnosis, and on what basis? In the wide range of human behavior, was there a boundary between sanity and insanity, or simply a continuum on which all people could locate themselves? And if Guiteau could be diagnosed as insane, what were the consequences of that condition: Was he responsible for his actions? Could he, by an exertion of willpower, have stopped himself from committing the horrible act of murder?

Beard had addressed that question in 1869, when he was defining and publicizing neurasthenia, a different illness entirely from insanity. Insanity, Beard maintained, was an organic disease of the brain. And just as a person could not be held responsible for contracting smallpox or scarlatina, so a person could not be held responsible for insanity, which often resulted from heredity. A person with smallpox could not, through an exertion of willpower, stop the course of the disease; a person with insanity could not stop the course of that illness either. Beard was called as an expert defense witness during the Guiteau trial, but was prevented from testifying on a legal technicality. Nevertheless, he repeatedly made known his conviction that Guiteau was, without a doubt, insane: he was a hereditary monomaniac, and therefore not responsible for the assassination of Garfield. Other physicians testifying for the defense upheld the same view, but the prosecution's expert medical witnesses disagreed vehemently, making an argument that the redoubtable editor E. L. Godkin had summarized in *The Atlantic Monthly* in September, while Garfield still struggled for his life. Godkin proposed that Guiteau was not insane, but merely a member of an erratic group, familiar surely to all of his readers, who are able to distinguish reality from fantasy and able, if intermittently, to work. "Their unsoundness and inability to succeed," Godkin wrote, "consists largely in a quality which is prominent in savages, but in them is ascribed not to insanity, but to imperfect development,—namely, want of tenacity of purpose." Guiteau was unstable, vain, selfish, and deluded, but until he shot the President, none of his actions identified him as a madman.[25]

Dissension in the medical community was widely publicized as the trial proceeded. If there was no scientific evidence to prove how the mind worked, how could Beard so adamantly claim authority? What did the dispute say about professional expertise? "The Doctor in the witness-box: this is a phenomenon of the present century," physician Charles L. Dana wrote in the *North American* Review, "and an evidence that science is expanding and human affairs becoming more complex." But instead of offering a consensus based on empirical evidence, the doctors called to testify in many cases, including Guiteau's, argued among themselves, exposing to nonscientists that evidence was open to seemingly endless interpretation. "[A]lthough science has now woven itself into every detail of human affairs," Dana complained, "it is a notable fact that the evidence of experts in medical sciences has continually lost ground, until today its inconsequential nature, and meager value in helping the ends of justice, are universally acknowledged."[26]

The dissension, moreover, contradicted the shared sense among the public that Guiteau, like any other human being, had the capacity for self-control. Beard's assertion seemed deterministic or materialistic, implying that people were at the mercy of heredity or biology. "The belief that there is a will, a volition, a force outside of or independent of the brain or the mind, a separate, distinct, special, isolated faculty or aggregation of faculties, is as baseless as witchcraft, astrology, alchemy, and spiritualism, and is as universal among philosophers and among the people as were all these delusions in the fourteenth century," Beard wrote. Instead, he argued, the will was merely "the coordinated action of all the faculties" of the physical body.[27]

Yet this assertion, by which Beard meant to liberate patients from blaming themselves for their maladies, was troubling to those who followed Guiteau's trial. Beard had devoted his career to inventing and publicizing categories of mental illness—neurasthenia and a host of what he called "borderline cases"—that could apply to many people. Individuals suffering from these borderline cases were able to function successfully within the community, did not need to be committed to asylums, and yet were beset by morbid fears. In their most virulent forms, borderline cases included agoraphobia, claustrophobia, and anthropophobia, or fear of men. One anthropophobic patient was so severely afflicted that she was afraid to keep her appointments with Beard. When the mind is disturbed to this extent, Beard claimed, responsibility is weakened but not obliterated. Yet this weakening of the will, this assertion that a patient could not achieve wellness by striving for it, made neurasthenic, depressed, anxious patients feel stripped of power and autonomy. The will as transcendent entity and willpower as a vital force were superstitions that many wanted to keep.

While Beard, a medical expert and scientist, protested against Guiteau's conviction and execution, newspapers and magazines reflected the public's desire to condemn Guiteau as a man who could have and should have been able to control himself. Despite Guiteau's bizarre behavior during his trial—he frequently would burst into childlike singing, for example—and despite his lawyer's efforts to have him declared legally insane, it took only slightly more than an hour for a jury to declare him guilty. The verdict generated "universal relief," showing, as one writer put it, "how wide-spread was the apprehension that the law had darkened counsel by words without knowledge. The great satisfaction lay, not in the assurance that the law had been successful over crime, but that common sense had been successful over law."[28] Common sense, shared perceptions among nonexperts who relied on their own experiences: this sense contained the wisdom and probity that must shape all decisions. Despite the efforts of reformers who saw the case as a judicial abomination, the public approved when Guiteau was hanged on June 30, 1882.

In the Guiteau case, commentators suggested that scientists not only failed to impart wisdom about ethics but could not be trusted even in physical matters. There was, after all, the problem of finding the bullet lodged in Garfield's body. After waiting to see how Garfield fared without their intervention, his physicians decided to probe for the bullet and, if possible, remove it. Since the X-ray had not yet been invented, they had to resort to manual probes, risking infection each time they entered Garfield's body. Still, it was their only recourse, and with trepidation, they proceeded. But the bullet was so deeply imbedded that they could not find it. The press reflected and heightened the public's distress, quoting physicians who advised against probing, and reporting a host of other possible remedies for the problem, offered by anyone, not necessarily from the medical community, with a new idea.

Finally, toward the end of the summer, Simon Newcomb, astronomer, psychical researcher, and one of the country's most noted scientists, came forth with a possible solution. He had been conducting experiments testing the effects of metal on electrified wire, finding that when metal was placed near the wire, he could hear a faint hum. It might be possible to use this process to locate the bullet, Newcomb thought, but he was pessimistic that the hum, faint as it was, would be heard. Alexander Graham Bell, reading about Newcomb's experiment, offered to apply his own talents to the problem. Perhaps, he thought, a telephone could amplify the sound. For several weeks, Bell and Newcomb worked to invent an instrument, testing it by hiding a bullet somewhere on their own bodies, and taking their device to a hospice for Civil War soldiers, where they attempted to find bullets lodged in veterans. When the instrument seemed reliable, they brought it to the White House.

Garfield, understandably, was fearful that if electricity passed over his body he could be electrocuted, even though he knew both Bell and Newcomb and trusted their expertise. Warily, he submitted to their electrical probe. But the results were inconclusive. Because Garfield was lying on a mattress of metal coils, which the two scientists did not realize, a hum was emitted everywhere that the electrified wand passed.

As the press reported the repeated failure, the public increasingly lost faith even in two reputable men of science. "Electricity came forward . . . with an ingenious scheme to discover the ball by some mystical metallic affinity," wrote Gail Hamilton in the *North American Review*. "The world was bidden to bend its ear and hearken to the hum and buzz of the obedient bullet responding to the summons of the marvelous machine. How it did hum and buzz! We heard it from Maine to California, and," Hamilton added sarcastically, "did obeisance to science." Science, accused of "vain-glorious boasting," had proven itself merely "a mewling and puking infant" with no right to present itself as the only route to knowledge, no right, for example, to promote itself as superior to religion as a path to truth. Instead, Hamilton suggested, science would be well "to give herself exclusively to sharpening her own eyes and strengthening her own muscle."[29]

The public's erosion of confidence in science, intensified by the Garfield case, recurred, just as sensationally, a few years later, focused again on a trial and its consequences, and again implicating the power of electricity.

PART III

ELECTROSTRIKES

CHAPTER 9

Live Wires

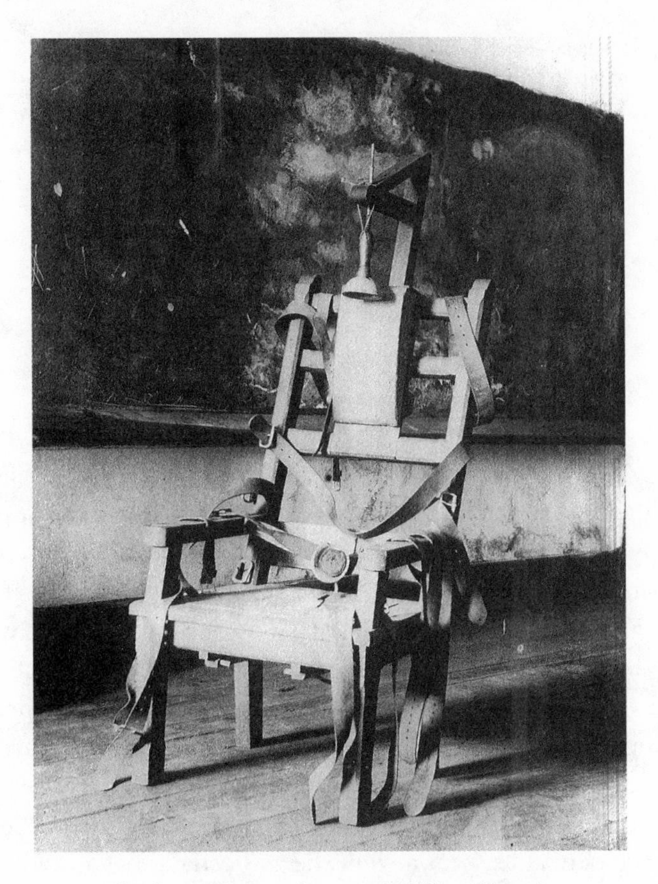

The electric chair used to execute William Kemmler at the Auburn Prison.

Collection of the Cayuga Museum

It was so terrible that the word fails to convey the idea.

The New York Times, August 7, 1890

In 1849, Alexandre Dumas père wrote a short story, "The Slap of Charlotte Corday," in which he considered the possibility that severed heads were capable of feeling and of communicating with the living. The story takes place in Paris just after the French Revolution when, the narrator tells us, "[t]here was an abundant harvest of blood" resulting from multiple beheadings every day. The narrator, trained in medicine but not a practicing physician, decides to test his conviction that "feeling is not entirely destroyed by the instrument of death"; since the brain is the "seat of feeling," and since the severed head contains blood, at least for a while, he was certain that a head not only could gnash its teeth in pain but also could speak. He devotes himself to experimenting on heads brought to him by cooperative executioners, and these grisly investigations prove his hypothesis—and disprove the notion that the guillotine, proposed as a more humane form of execution than other means, fulfilled its benevolent purpose.[1]

Yet thirty years later, on the other side of the Atlantic, the guillotine resurfaced as a possible alternative to the inhumane gallows. By the 1880s most states had outlawed public hangings, and an anti-gallows movement had taken firm hold. The American public was revolted by a method of execution that simply took too long. The condemned were left dangling—gasping, limbs twitching convulsively, eyes rolling with pain—for half an hour or more, and even then sometimes were cut down alive. Hanging was barbaric and unusually cruel; surely, some Americans speculated, an advanced civilization could devise a more enlightened way to kill.

The state of New York took the lead, spurred by a Buffalo dentist, Alfred Porter Southwick, who in 1881 had seen an accidental electrocution from a live wire and marveled at the victim's apparently instantaneous death. The enterprising Southwick began to conduct experiments on animals, and after a few years became convinced that electricity offered a solution. In 1885, his friend state senator Daniel H. McMillan introduced a resolution to appoint an investigative commission charged with suggesting a new form of capital punishment, a resolution that Governor David Hill amply supported. "The present mode of executing criminals has come down to us from the Dark Ages," Hill told the state legislature, "and it may well be questioned whether the science of the present day cannot provide a means for taking the life of such as are condemned to die in a less barbarous manner."[2] Among those less barbarous possibilities was death by electrocution.

Electricity, however, was not the only option. The commissioners—Southwick, and lawyers Elbridge T. Gerry and Matthew Hale, assisted by nine researchers—probed extensively into the history of execution. They consulted hundreds of sources, including books on criminal law, encyclopedias, and histories; they contacted some seven hundred authorities; they considered all methods of death and torture from all cultures; and finally they identified a menu of methods that fit, they said, "the requirements of humanity": the garrotte, the guillotine, shooting, lethal injection, and electricity.

They then designed a questionnaire polling opinion about these methods and sent it to hundreds of prominent New Yorkers, primarily judges, district attorneys, sheriffs, and physicians. By January 1888, the commission submitted its report based on two hundred replies. Eighty, they noted, recommended no change: After all, many reasoned, what did it matter if condemned murderers suffered on the gallows? Others, however, disagreed, but found most alternatives as undesirable as hanging. The guillotine seemed most merciful for its speed, but repulsive because of "the fatal chop, the raw neck, the sprouting blood"; moreover, it had unpleasant associations with the horrors of the French Revolution. The garrotte, while having the advantage of "celerity and certainty," was, in effect, no less distressing than the guillotine in terms of bloodiness; and it, too, had negative political associations, this time with Spain and its colonial policies. Shooting might be appropriate in military executions, where competent marksmen were available, but was too "bloody in its character and effects" to be acceptable in civil cases. Lethal injection with prussic acid seemed eminently humane, but responding physicians objected to the use of hypodermic needles, which aroused enough suspicion among patients, to promote death. One physician suggested a gas chamber, into which carbon dioxide or monoxide would be pumped at night, while the convict slept.[3] Eight responders were undecided. And so it was left to electricity—variously called electromorsis, electricide, electricission, electrostrike, or ampermort—which received a small majority of votes.

The commission gave only a passing glance to the question of whether capital punishment should be enacted at all, conceding, however, that no death penalty ever had served as a deterrent to capital crimes. Nevertheless, they did believe that electricity offered a quick, painless, and cost-effective means of death. Since electrical currents were one hundred times quicker than nerve current, they claimed, electrocution would insure that the convict was unconscious within seconds and would feel no pain from the lethal shock. They recommended, in short: "The punishment of death must in every case be inflicted by causing to pass through the body of the convict a current of electricity of sufficient intensity to cause death, and the application of such current must be continued until such convict is dead."[4]

The commissioners' report set off debate among law officials, physicians, and, not least, those with financial interest in the future of electrical applications. Electrocution became an issue as much about money as ethics, and about a struggle for power more insidious than electrical current.

Since many physicians administered electrotherapy as part of their arsenal against illness, they were wary of sanctioning electrocution as the ultimate punishment. At a meeting of the Society of Medical Jurisprudence, the notable Dr. William Hammond argued, instead, in favor of a new, "scientifically adjusted" method of strangulation that would be both "effectual and merciful... The subject first feels great heat in the head, bright lights dance before his eyes, there is a tingling all over the body, roaring sounds in the ears, sometimes ravishing sounds of music, a feeling of heavy weights to the feet, then—insensibility. There is no testimony of pain that I have discovered." Of course his colleagues wondered how there was any testimony at all, from the dead. Dr. Emil Brill, for one, suggested that there was no physiological basis for Hammond's conjectures; another physician objected to any mercy shown the convict and wanted "to let the punishment fit the crime, no matter if inquisitorial torture was necessary." Several physicians believed "there was nothing like the guillotine"; and only one suggested that the murderer might be compelled "to support those whom he had robbed of a protector."[5]

Besides objections from physicians, the electrical community voiced anxious concerns. Any hint of the lethal potential of electricity could undermine efforts to market the new form of energy. Yet among the Edison camp, some believed that electrocution could serve their own needs, bolstering the argument for direct current rather than the alternating current promoted by Nikola Tesla and George Westinghouse, who had become Edison's most tenacious business rivals. But Edison was determined to prevail over Westinghouse by inspiring fear: high current, he insisted, meant danger.

With little—or sometimes no—understanding of what electricity was, where it came from, or how it energized lights and motors, there seemed to be no way for people to evaluate what they read. For some consumers, the arguments between Edison and Westinghouse only fueled a more generalized fear of electricity itself. The physician Peter H. Van Der Weyde tried to put the debate into historical perspective: "The opposition of the alarmists who predict terrible disasters from the introduction of alternate electric currents," he told participants at a convention of the National Electric Light Association in August 1888, "reminds me forcibly of the predictions made less than one century ago, when the world was warned against the introduction of illuminating gas and railroad trains. About gas it was prophesied that it would blow up cities, or destroy them by universal conflagrations, and that railroads would cause the indiscriminate and wholesale slaughter of the reckless individual who dared to tempt Providence, by trying to travel with the enormous velocity of twenty miles an hour."[6]

IGNOBLE PURPOSE

On June 4, 1888, barely five months after the commission submitted its report, Governor Hill signed into law the Electrical Execution Act, abolishing hanging and substituting death by electrocution, to take effect for all murders committed after January 1, 1889. If electrocution seemed a more civilized alternative to hanging, still there were significant questions unanswered. How much current would be needed to kill a criminal? What kind of dynamo would be suitable for the task? How long should the current surge through a body? How would an individual be attached to electrical wires? To address these crucial issues, Hill created the Electrical Death Commission, whose task it was to discover just what kind of and how much current was needed to kill a human being, to oversee testing, and, most crucially, to invent the electrocution device. The commission was eminent: Southwick, of course, champion of electrocution; Dr. Carlos F. MacDonald of the State Asylum for the Insane; General Austin Lathrop, superintendent of state prisons; Alphonso D. Rockwell, electrotherapist; and Harold P. Brown, a self-proclaimed electrical expert, affiliated with Edison, determined to make sure that electrocutions would be caused by alternating current.

From the first, Edison took a prominent, if apparently benignant, role in the proceedings: he set aside a building in his laboratory compound for the commission's use, where dogs, calves, horses, and cows were electrocuted, dozens in all, by alternating current. The commission had no lack of animals provided by the boys of Orange, New Jersey, who took advantage of a new way to make easy money. Although the boys apparently showed no regret about leading the animals to their death, a *New York Times* reporter who witnessed the animal executions betrayed the anxieties of all involved. The room, the reporter noted, "was fitted up with coils of wire, clusters of lamps, and sundry and divers instruments which inspired a wholesome awe in the hearts of the uninitiated, and led them to keep at a respectful distance." Into this forbidding laboratory, Edison's technicians led the animals, one by one.

If the dogs were calm as they entered the execution chamber, the attachment of wires to legs and head "appeared to awaken suspicion," and by the time the matrix of wires, plates, and sponges was in place, the dogs were "apparently somewhat anxious as to future events." So were investigators, who backed away gingerly as each experiment was about to begin, not knowing what to expect.

The experiments were enacted to prove that alternating current—with repeated surges of high voltage—killed "by a series of knock-down blows," quickly and apparently painlessly; and, furthermore, the experiments would determine just how much voltage was needed to cause death and how long the current must be administered. To the *Times* reporter, death did seem to occur instantly. A black, eighty-seven-pound Newfoundland mongrel died within ten seconds from a current of five hundred volts. First its limbs stiffened as if it were about to pounce forward, but, said the reporter, there was "no sound, no convulsion, no appearance of pain at all," and when the last wire was taken off, the dog collapsed in a heap. The calves, more skittish than the dogs, required eight hundred volts because of their size; and twenty-five seconds of a thousand volts killed one sad-eyed horse.

The results, to all observers, were conclusive: alternating current was a "sure agent to take the place of the hangman's rope." But still, no one could tell how much voltage should be used to kill a man, how long the current should be kept on, nor even what the death machine should look like. Several commission members toyed with designs: Southwick favored a wooden chair, chillingly like a dentist's chair, in which the prisoner would sit with electrodes attached to his head through a small cap, to his spinal cord, and to his hands and feet. Yet by December 1888, no firm design had been approved, and the law would take effect in weeks.

The law also was threatened by growing anti-electrocution sentiment among electricians. Just as physicians did not want to be associated with death by lethal injection, the electrical community balked at flaring publicity about the lethal potential of electricity. In January 1889, at an electrical convention in Chicago, manufacturers adopted a resolution declaring "that no company should allow a current in its control to be used for the 'ignoble purpose' of executing murderers." But, although the decision was widely publicized in the press, it hardly affected the process: electrical lighting generators, after all, could be purchased anywhere to be used in any way the purchaser chose. After Governor Hill allocated $10,000 for electrocution equipment, Harold Brown closed a deal with the state for three Westinghouse lighting dynamos—bound for Auburn, Clinton, and Sing Sing—for a total cost of $8,000. The deal stunned Westinghouse. So did Brown's suggestion for a new term for electrocution: to *Westinghouse*.

HOPELESSLY DEADLY

William Kemmler, aka John Hart, a twenty-nine-year-old vegetable seller, the illiterate son of German immigrants, and apparently an alcoholic, lived in Buffalo with his twenty-four-year-old lover Tillie Ziegler and her young daughter, Emma. Both Kemmler and Ziegler were married to other people and had eloped from their native Philadelphia to elude their spouses. But their life together was volatile and violent: Ziegler threatened to throw Kemmler out and send him back to his wife; Kemmler, especially when he was drinking, fell into jealous rages. On March 28, 1889, he attacked Ziegler with an ax, leaving twenty-six gashes in her skull. Barely alive when the police came, she died the next day.

Kemmler was arrested and, after a four-day trial, convicted and condemned to receive "a current of electricity of sufficient intensity to cause death." The first electrocution was scheduled for June 24. But on May 18, when the sentence was pronounced, Charles H. Durston, warden at Auburn Prison, admitted that nothing was yet prepared: not the electrical chair, not even the cell where, according to the new law, Kemmler was to be kept in solitary confinement. In fact, no one was certain how much current would be sufficient to cause death. To help solve that question, prison superintendent Austin Lathrop received a query from a Philadelphia man "down on his luck" who offered himself for experimentation in exchange for $5,000 for his wife.[7]

Now that New York State had a convict ready for electrocution, Westinghouse intensified his efforts to prevent the use of alternating current. W. Bourke Cockran burst onto the scene, a lawyer whose sudden interest in Kemmler focused on the constitutionality of the new law. Electrocution, he objected, was cruel and unusual punishment, and he asked that Kemmler's execution be stayed until that argument could be made in court. Cockran's plea was granted, and on July 9, 1889, the case of William Kemmler against Charles H. Durston, warden of Auburn Prison, officially opened in Cockran's Manhattan law offices, with lawyer Tracy Becker assigned as referee. Harold Brown was the first witness, testifying vigorously to the efficiency and painlessness of electrocution, proven by the experiments on animals the previous year. Nevertheless, Brown became rattled when Cockran pressed him about the effects of electricity on humans. Didn't the experiments allow Brown only to draw inferences? Cockran asked. Wasn't it true that he had never ascertained how much current was needed to render a man unconscious? Wasn't it true, furthermore, that some individuals had survived intense shocks and even lightning strikes? Although Brown insisted that alternating current "was by its very nature hopelessly deadly," Cockran so forcefully badgered him with theoretical questions that Brown became confused, finally claiming that he was an expert only on "commercial electricity" and not on electrical science. After the first day of hearings, it appeared that Westinghouse had won a round.

After Brown's testimony concluded on the next day, Cockran heard from Westinghouse electrician Franklin L. Pope. Cockran directed Pope to the controversial issue of the Wheatstone bridge, a device, according to Brown, that would measure the electrical resistance of an individual and therefore help to ensure that sufficient current was administered to cause instant death. The Wheatstone bridge was unreliable, Pope contested, because too many variables—dirt, perspiration, thickness of skin—affected an individual's resistance. "With the Wheatstone bridge you get a result," Pope said, "but there is no means of determining the accuracy of that result, nor do I think it possible to do so." Pope also was skeptical that Brown's animal experiments could be applied to humans; because resistance could not be determined accurately, the same voltage that might kill one person would burn and scald another.

Pope kept his calm certainty even under cross-examination, when he admitted that the Westinghouse Company objected to electrical executions because they would incite the public's fears of electricity for commercial use. "Has the company engaged counsel to urge this objection?" Pope was asked. "Not to my knowledge," he replied with equanimity while Cockran, according to a *Times* reporter, "suddenly became intensely interested in the architecture of the ceiling."[8]

The hearings continued, with witness after witness testifying to uncertainty. "When we deal with electricity we are not sure of anything," said Daniel L. Gibbens, the youngest member of the Board of Electrical Control. "We have no means of knowing whether a man who has been killed by electricity was killed instantly or not, or painlessly. We don't know how it kills; we simply know that the man is dead."[9]

On July 12, instead of convening in Cockran's office, the investigating team went to Edison's workshop and tested their own electrical resistance with the famous Wheatstone bridge: the results, as Pope had predicted, varied so greatly, even in the same person, that it seemed impossible to know how much current would be needed to insure death. Still, when Elbridge Gerry, who had been vacationing on his yacht and not eager to return, finally appeared as a witness, he defended the commission's findings. Unlike some previous witnesses, Gerry, a lawyer, was unintimidated by Cockran, and their banter entertained the largest crowd yet assembled for the hearings. Gerry explained in great detail the exhaustive research carried out by the commission on which he, Hale, and Southwick served. He conceded that he himself was not an electrician, but, he pointed out, neither was Cockran: "my knowledge just about equals yours," Gerry argued, "and the less you know about it the better you seem to cross-examine."[10]

Gerry's testimony in favor of electrocution was followed by a roster of physicians who took the opposite view and of men who had received severe shocks and lived to tell about it in dramatic detail. Rockwell countered those reports with his observations of the Electrical Death Commission's experiments and his own knowledge of human physiology, holding firmly to his convictions even under Cockran's patronizing questions. Yet despite Rockwell's testimony, Westinghouse seemed to be making a strong case—that is, until July 23.

"TESTIMONY OF THE WIZARD," read the headlines in the *Times*. Edison, accompanied by his chief electrician Arthur Kennelly, arrived in Cockran's office on Tuesday morning, July 23, smiling convivially at his opponent. Deputy Attorney General Poste began, shouting his questions so loudly that they resounded in the meeting room. Edison explained the difference between alternating and direct current, testified to his repeated experiments in determining the body's resistance levels, and maintained that continuous current simply does not have much effect on nerves. When Poste finished, Edison, still smiling, dragged his chair across the room and seated himself with his good ear turned, expectantly, to Cockran's mouth.

If Cockran hoped to demolish his star witness, he failed. Instead, Edison so completely frustrated Cockran that the lawyer's face burned as red as a boiled lobster, the *Times* noted. Edison did agree, however, to repeat his experiments to determine the body's resistance levels, and did so soon after he returned to his laboratory on Tuesday afternoon. Of the 226 men that he tested and retested, he reported, results were remarkably consistent each time.[11]

Edison was the star witness, but not the last. By the time the hearings had concluded in Manhattan on July 25, more physicians, electricians, and experimenters testified; Cockran still maintained that resistance was too variable to be dependable; and the only firm result of the three weeks seems to have been the many friendships forged among the participants. Everyone shook hands warmly before departing, some for Buffalo, where hearings continued.

As Kemmler's appeals continued, a committee appointed by Superintendent of Prisons Austin Lathrop arrived in Auburn on New Year's Eve of 1889 to test the chair and the Westinghouse dynamo that electrified it. Among the members were Rockwell, Carlos MacDonald, and Professor Landy of the Columbia College School of Mines, all charged with submitting written reports on their findings at Auburn and then at Clinton Prison. At the first try, the dynamo broke down, and while waiting for it to be repaired, the observers had a chance to tour the prison, where they were treated to an impromptu entertainment of vocal imitations of a bugle and cornet by a talented black convict. They were so impressed that they took up a collection for the prisoner.

The afternoon's events proved more dramatic: with the dynamo finally fixed, the observers watched as an old horse and four-week-old calf were electrocuted. The horse died in less than half a minute, with "no vulgar demonstration" of pain, the calf in less than ten seconds. The committee members, whom reporters characterized as "gleeful" about the results, proceeded to Clinton Prison, where a young bull's electrocution confirmed the Auburn tests. At both sites, however, the committee found that the dynamos themselves posed problems; delays occurred because various pulleys, wires, and belts needed to be adjusted and aligned.

As the concurrent appeals and testing made their way into the press, it became increasingly clear that justice for Kemmler was not the issue. "The corporation the apparatus of which had been chosen for the infliction of the penalty had an interest in making it appear that persons subjected to the current generated would not die, although the evidence that they had died was conclusive and the probability that they would die was simply overwhelming," the *Times* complained in March, just after the court of appeals upheld the constitutionality of New York's law. "The judicial scandal was that the time of the courts was taken up in trying to persuade the public of the harmlessness of their machine."[12] But the case did not end there: Kemmler's—or Westinghouse's—lawyers took it to the U.S. Supreme Court, where on May 19, 1890, Chief Justice Fuller ended the debate: "Punishments are cruel," he wrote, "when they involve torture or a lingering death; but the punishment of death is not cruel within the meaning of that word as used by the constitution. It implies there something inhuman and barbarous, something more than the mere extinguishment of life."[13]

Once again, Kemmler appeared in court for sentencing, and a new execution date was set for August 4. Although Westinghouse made one last effort to repossess its dynamos from the three prisons, the company finally gave up. They, and Kemmler, realized that they were doomed.

THE GREAT EXPERIMENT

Throughout the summer, the town of Auburn buzzed with rumors about "the great experiment"; even as late as August 3, some doubted that the event would occur at all; there were reports that Kemmler, after a year in solitary confinement, had gone crazy, and should be judged insane and have his death sentence commuted. But Warden Durston, repeatedly interviewed, denied those rumors; Kemmler, he insisted, was "no more an idiot today" than when he first arrived. The same could not be said for Durston, who seemed to grow more agitated, irritable, and short-tempered day by day. His relief at each stay of execution and his nervousness as August 4 approached fueled yet another rumor: that he had traveled to Manhattan to formally present an insanity plea on Kemmler's behalf.

On Monday, August 4, Durston sent out invitations to the twenty-seven men who would witness Kemmler's death, summoning them to the prison at seven o'clock on Wednesday evening. The crowds milling outside the prison walls took this to mean that Kemmler would be electrocuted early the next morning. The execution had to occur during daylight so the pathologists could conduct an autopsy immediately: despite the special dynamo for the death chair, Auburn was not wired for electric lights.

Durston also kept Kemmler informed about the plans, a part of his job that he found especially abhorrent. Durston clearly had developed a fondness and respect for his prisoner, whose simplicity and calmness impressed him. He saw that since the final sentencing, Kemmler had changed—not becoming mad, as some townspeople speculated, but descending into depression. Although the press portrayed Kemmler as having a "strange nature... incapable of thinking and feeling as other men feel," Durston knew otherwise.[14] On Monday afternoon, he allowed Kemmler some company, a fellow convict, who brought his banjo into Kemmler's cell and entertained him for a few hours. It was the only time that Kemmler roused from apathy. Even a letter from his brother, the first he had received from any family member, failed to elicit any display of emotion.

Besides the twelve jurors who condemned Kemmler and the district attorney and sheriff of the Buffalo county where Kemmler was convicted, the witnesses included Carlos MacDonald, charged with writing a report for the governor; Southwick; Daniel McMillan, who had introduced the Electrical Execution bill; Tracy Becker; several physicians, including a tracheotomy specialist, in case Kemmler needed to be revived; two ministers; a representative of the Dunlap Cable News Company; and a reporter from the United Press and one from the Associated Press. Hundreds of other reporters waited outside of the prison, and a special corps of telegraph operators sat at tables, ready to send their stories across the nation and to the world.

The team of physicians had come not only to witness the execution but to perform the long-awaited autopsy. This electrocution, scientists knew, afforded an unparalleled opportunity for discovery. In most cases of electrocution, a delay between the death and the autopsy compromised any findings; here, at last, pathologists could examine every part of the body immediately to determine the effects of an electrical current. "The blood, the organs, all the nerve centres, the muscles, and every part of the remains will be studied as a historian would study a rare manuscript and the action of the mysterious current on each carefully noted," the *Times* commented. The autopsy "will be so thorough that it will pass down to history in medical annals."[15]

Despite their long anticipation of the execution, on Tuesday evening, when the townspeople of Auburn saw the witnesses start in a procession from their hotel to the prison, they felt unaccountably stunned. By the time the witnesses entered the prison, a huge throng had gathered to follow them; when the iron gates closed behind the procession, the crowd remained shut out, speaking only in whispers, as if, the *Times* reported, "a feeling of awe had come over them." After half an hour, a rumor circulated that the execution had been carried out, and the crowd began to disperse; but soon they learned that the witnesses had assembled merely for an informational briefing, and the townspeople, disappointed but relieved, followed them back to the hotel. Nothing would happen, they found out, until dawn the next morning.

Even before dawn, townspeople began to gather outside of the prison; by five o'clock there were more than five hundred; an hour later, it was hardly possible to break through the crowd that stood anxiously under the clear blue sky of a perfect summer morning. Some young men climbed nearby trees and telegraph poles. Durston, meanwhile, went to Kemmler's cell to read the death warrant. "All right, I am ready," Kemmler replied steadily. It seems that he was: he shared a hearty breakfast with two ministers and knelt with them as they prayed. He cheerfully agreed to have his hair cut, much less nervous, apparently, than the man who volunteered for the job—Joe Velling, the warden at the Buffalo prison where Kemmler began his ordeal. "They say I am afraid to die, but they will find that I ain't," Kemmler told Velling. "I want you to stay right by me, Joe, and see me through this thing and I will promise you that I won't make any trouble."

If Kemmler was ready to die, his executioners were not ready to enact what suddenly seemed to all of them an awful deed. "Gentlemen," Kemmler said after being introduced to the group of witnesses, "I wish you all good luck. I believe I am going to a good place, and I am ready to go." The death chamber contained only the chair and seats for witnesses; the Westinghouse dynamo was housed in another wing, about a thousand feet away, with wires running out of a window, along the roof, and into the execution area. Those wires in turn were connected to a switchboard located between the dynamo room and the death chamber. A system of signals had been devised between Durston, in the death chamber with Kemmler, and the two electricians who manned the switchboard and dynamo: one bell to turn the current on or off; two to increase it; three to reduce it.

With the machinery well hidden from him, Kemmler settled into the wooden chair, then, at Durston's request, rose to have his pants and shirt slit to allow contact between his skin and the electrodes. As Durston began to arrange the wires, Kemmler turned to him. "Now take your time and do it all right, Warden," he said. "There is no rush. I don't want to take any chances on this thing, you know." Carefully Durston adjusted the headpiece and buckled the eleven straps that bound Kemmler to the chair. Sunlight streamed through the chamber window onto Kemmler's cheek. "Good-bye, William," Durston said softly. And with that, the witnesses heard a soft click: a lever had been pressed, the process began.

"Words will not keep pace with what followed," the *Times* reported. The witnesses saw Kemmler stiffen in his chair, his skin changing from pale to dark red. His right hand convulsed so violently that the nail on the index finger cut through the palm. His face twitched; his whole body convulsed and strained violently against the straps.

Although the current was supposed to have continued for twenty seconds, the witnesses could bear no more. "Stop!" many shouted, and Durston relayed that order to the switchboard: seventeen seconds had passed. "He's dead," announced the witnesses who dared to approach the chair; but one physician, examining the skin, noticed blood dripping from the cut in Kemmler's palm: if blood was still dripping, the heart was still beating. "Turn on the current! Turn on the current!" he shouted, and others echoed his cry. But the current could not be turned on instantly because the dynamo, which had been completely shut off, needed to build up power. The witnesses watched in horror: they believed they could see Kemmler breathing. Some could hear a gurgling from his throat. One reporter fainted; Arthur Quimby, a district attorney with a reputedly hard heart, ran from the room groaning in dismay. Finally the current again coursed through Kemmler's body, but this time the witnesses heard sharp snaps from the wires, and blood from ruptured vessels broke out on Kemmler's face. Then there was the smell, a terrible smell of burning flesh and hair, and they saw that hair under the headpiece and skin at the base of the spine were singed. How long was the current left on for the second time? No one was sure: two minutes, four. By the time Durston signaled again, many witnesses were nauseated, some were in tears, and all, the *Times* reported, left the death chamber "as miserable, as weak-kneed a lot of men as can be imagined... They all seemed to act as though they felt that they had taken part in a scene that would be told to the world as a public shame, as a legal crime."

No one rushed to dissect Kemmler's body. More than half an hour passed before physicians even could approach the death chamber again, and when they did, they agreed to leave the body for three hours before performing the autopsy; they wanted no suggestion that Kemmler "died under the scalpel, and not from electric current."[16]

The next day's newspapers were nearly unanimous in their condemnation of the electrocution. "Apparently," wrote the *New York World*, "the man died in agony, by slow torture." "[T]he wretch was actually tortured to death with a refinement of cruelty that was unequaled in the dark ages," added the Chicago *Evening Post*. "It may be taken for granted that public sentiment in New York will tolerate no further essays in this new experimental science of man-killing by electricity, pending the time when the law can be repealed in the name of the State's dignity and of the enlightened humanity of the nineteenth century," the New York *Sun* concluded. The Rochester *Morning Herald* agreed, calling "in the name of humanity and for the sake of the morals as well as the sensibilities of the people of this state" that the first electrocution would be the last. Other papers likened the event to burning at the stake and to the tortures of the Inquisition. "Now," wrote the Elmira *Telegram*, "it is the turn of the people of common sense and humane ideas. There must be no more executions by electricity in the Empire state."[17]

But if the people of common sense rose up against electrocution, the experts supported it. Seconds after the first application of electricity to Kemmler, Southwick proclaimed success, and he never wavered. In his official report, Carlos MacDonald acknowledged "the wide publication of unofficial reports... and the efforts which have been made to proclaim it a failure, and to invest it with an air of repulsion, brutality and horror"; yet he, too, "confidently believed" that he had seen a "successful experiment" and "a step in the direction of a higher civilization." Yes, the second application of the current was left on too long, and yes, there had been "momentary scorching"; but the burning occurred, he insisted, after the body was already dead. He was certain that "the intent and purpose of the law, to effect sudden and painless death in the execution of criminals, had been completely and successfully carried out."[18]

This official report outweighed reports of the public's horror. In the next few years, there were many occasions to test

MacDonald's conclusions, as criminals were electrocuted at Sing Sing, Clinton, and again at Auburn. Witnesses were revolted by the procedure that, more often than not, required reapplication of the current; at Sing Sing, in 1900, a convict was administered five shocks before he was pronounced dead; the death chamber, more often than not, filled with the nauseating smell of burned flesh and hair. "Although I have seen many dreadful things," recalled physician Allan Hamilton, witness to several Sing Sing executions, "I don't think any other has ever raised my gorge as this had done, and for weeks my dreams were filled with details of that half hour." If other executions were carried out as "solemn affairs," in electrocutions, it seemed "that every one thought more of the success of the procedure than that a human being, no matter how wicked, was being sent out of the world with so short a shrift." A physician who served as Boston's medical examiner confessed to Hamilton that after being present at many electrocutions, he believed it "brutal and unscientific" and hoped that it would soon be abolished.[19]

DEFYING DEATH

Criminal electrocution seems an unlikely theme for fiction, but Arthur Conan Doyle, a thirty-one-year-old physician just about to give up his medical career for writing, read the press's graphic reports of Kemmler's execution, and he saw in the event a chance to write about something that deeply interested him: vitalism. Trained at the Edinburgh Medical School, from which he graduated in 1881, Conan Doyle was experienced in the medical application of electrotherapy. Yet in his fiction, he reveals a certain skepticism about electricity's apparent power to cure and vitalize, a power, he implies, that comes as much from a patient's desires as from a battery. In the short story

"Crabbe's Practice," for example, the friend of a struggling physician pretends to drown so that he can be brought back to life by galvanic current, gaining notoriety—and patients—for the doctor. "The Great Keinplatz Experiment" pokes fun at both electrotherapy and mesmerism. Not surprisingly, Conan Doyle saw in New York's great experiment another chance to satirize the public's naïve and contradictory beliefs about electricity. Given his attitude toward vitalism, perhaps it is more surprising that Conan Doyle spent his last years as an ardent and outspoken proponent of spiritualism.

In his fictional western town of Los Amigos, the citizens, hearing troubling reports about the execution in New York, decided that they could enact one with much more success: they simply would apply a charge six times greater than that used on Kemmler. Soon enough they find a criminal, Duncan Warner, and proceed to condemn him to death. The townspeople are enthusiastic, except for Peter Stulpnagel, a German inventor, who quietly cautions against such a large dose of electricity. It would have, he says inexplicably, the opposite effect. But Stulpnagel persuades no one.

With great anticipation, the executioner gives the huge electrical jolt to Warner—but Warner is not even slightly injured; in fact, the shock does nothing except turn his hair white. Amazed, the executioner gives him another jolt, and this time his hair falls out. Clearly electrocution will not work on this criminal, the townspeople are puzzled to learn; and so they resort to their former, reliable method of hanging. But after dangling on the gallows for half an hour, Warner is still very much alive; angry and embarrassed, they shoot him, but still the resilient Warner doesn't die. The town, of course, is mystified. Only Stulpnagel can explain the mystery of Warner's immortality: "What you have done with your electricity," he explains, "is that you have increased this man's vitality, until he can defy death for centuries."[20]

Those who walked each day beneath a tangle of wires, fearful that a gust of wind would swipe a live wire across their face; whose town officers were debating, noisily and publicly, the advantages and dangers of buying an electrical system; certainly those who had followed Kemmler's plight: all were concerned less with electricity's potential to insure eternal life than with its likelihood to cause imminent death. Men and women who worked in wired offices commonly felt a mild shock, and even saw sparks when turning on a lamp or touching a switch—a shock that portended danger. Accounts of accidental electrocutions warned that even innocent bystanders could be felled by the force. In London, several devastating explosions occurred when the ground became electrified through faulty insulation of wires and when the conduits for electric wires accumulated explosive gas from sewers or gas mains. In Providence, Rhode Island, electric wires that followed a gas main caused a huge explosion affecting houses and buildings for thirty feet, blowing out windows, spewing shards of wood and metal.[21] Electricians especially were vulnerable to electrocution; newspaper reports offered disturbing details of the accidental deaths of young linemen, felled in the course of their jobs. Expertise did not mitigate the danger.

Though electricity was invisible apart from occasional sparks, it was hardly a silent force. Wires crackled, hummed, and buzzed. Moreover, when electric wires were strung next to telephone wires, they often set off the telephone system's alarm bells, burned out instruments, caused false calls, and interrupted conversations. Besides warring with each other over the use of direct or alternating current, electrical companies staged an acrimonious public debate with telephone companies.

As the century drew to a close, the public encountered repeated and disturbing dissonance between vast claims about the positive role that electricity would play in the future and evidence, such as criminal and accidental electrocution, of its tangible negative effects. Electrical technology would improve daily life, exert a powerful force for moral good, and create wealth and ease for all. Or the technology would fail again and again, cause death and destruction, and upset the balance of nature. The incandescent bulb would illuminate gently; the glare would cause blindness. Electricity would be affordable by all; electricity would cause a great rift between the social classes. Electricity would energize; it would deplete human vitality.

And still, most people had little access to the technology in their homes. In 1891, the National Electro-Therapeutic and Alarm Company advertised a nifty device that combined an electric doorbell, fire and burglar alarm, and medical apparatus, all contained within a single wooden box and selling for a mere $15. Except for electric lighting in public spaces, at work, and in those few homes that were wired, this device represented most of what was available to consumers at the time. Appliances were few: electrical sewing machines ran on six-volt batteries, not as reliable as a treadle machine operated by the sewer's muscle power. Edison had improved his phonograph, but when Mark Twain bought one to dictate his writing, he found the recording process "so awkward for me and so irritating that I not only curse and swear all the time I am dictating, but am impatient and dissatisfied because God has given me only one tongue to curse and swear with."[22]

Yet if the press was to be believed, soon electricity would be everywhere, for everyone, meeting needs that the public did not yet realize it felt. Writers such as Arthur Kennelly, an electrical engineer who joined Edison's staff in 1887, disseminated the news that "the electric Ariel" soon would cross the threshold into all homes. In "Electricity in the Household,"[23] published in *Scribner's Monthly* in 1890, Kennelly offered details and illustrations of how life would change: water pumps and elevators, phonographs and burglar alarms, coolers and heaters, all would be powered by electricity coursing within the walls of the house. New applications for the force were sure to charm consumers: in the dining room, for example, a miniature electric train, carrying food, would travel around the table on an ornamental trestle, disappearing tastefully behind a shutter in the wall. In the conservatory, tiny bulbs would be installed on beams, giving the effect of a hundred twinkling stars. In the study, an electric fan would provide comfort for a gentleman impelled to wear a suit jacket and vest in all seasons. "Ignorant as we are of the real nature of this marvelous agent," Kennelly wrote, "we know at least that electricity implies power; all the evidence by which we are rendered sensible of its presence are manifestations of energy." Consumers, Kennelly implied, would enhance their own power to ensure their family's security, provide for their pleasure, and direct their servants, as they wielded the force. This electrified future was not so distant, Kennelly assured his readers; soon electrical appliances would be considered not as extravagances but as necessary as tobacco: an irresistible addiction. As Kennelly well knew, plans already were under way to give his readers, and millions of other potential consumers, a vision of glorious and seductive electrical innovations: the World's Columbian Exposition, to open on the four hundredth anniversary of Columbus's landing, was to be the greatest electrical display of all time. Westinghouse, unharmed by Edison's efforts to discredit him, won the coveted commission to wire the fair; Edison's companies planned major exhibitions. By the time the exposition closed, however, the country had sunk into depression, and, more and more, the public encountered bleak and unsettling visions of an electrified future.

CHAPTER 10

Magical Keys

Interior of the Electricity Building at night, World's Columbian Exposition, from Glimpses of the World's Fair.

What is electricity? We do not know, and for practical purposes it is not necessary that we should know. We are only concerned in what its properties are—how we can make it obedient to our will.

Electricity in Homes, 1889

At noon on May 1, 1893, President Grover Cleveland stood before a cheering throng of more than a hundred thousand people and pressed a gold-and-ivory key, much like the key on a telegraph, to set in motion the greatest show on earth: a huge, white city, sparkling under the midday sun, a display, Cleveland declared, of "the stupendous results of American enterprise and activity." The World's Columbian Exposition, commemorating the fourth centenary of Columbus's landfall, was America's debut onto the international stage, testimony to its vigor, self-reliance, and ingenuity; testimony also, according to its promoters, to the triumph of democracy. "We have not only built these splendid edifices," Cleveland told the crowd, "we have also built the magnificent fabric of a popular Government, whose grand proportions are seen throughout the world. We have made and here gathered together objects of use and beauty, the product of American skill and invention." More important, he added, "We have also made men who rule themselves."[1]

Perhaps there was no more appropriate city than Chicago to represent American independence and self-rule. Described as a "dirt pile" and "muckheap" by one native son,[2] Chicago had wielded its considerable financial interests to win the lucrative prize of hosting the exposition, competing against New York, Washington, Minneapolis, and St. Louis. To gain the confidence of Congress, the city promised that visitors, even women traveling alone, would be safe walking its streets and sleeping in its hotels. Still, despite the promises and Chicago's obvious energy and ambition, the choice surprised Easterners, who feared that the "putative Philistines of the New World" could not help but produce a circus of cheap vulgarity.[3] The Chicago Directory, though, was not left to plan the exposition alone, but was under the guidance of a national commission, consisting of two representatives from each state and territory. The result was predictably contentious, although in the end, a notable group of leaders from the arts and architecture, mostly from the East, emerged to put their stamp on the largest exposition the world had ever seen.

Frederick Law Olmsted chose Jackson Park, a morass along Lake Michigan, for the site, and he designed the landscaping: expansive plazas, wide promenades, sinuous canals and lagoons, and a mile-long midway that served as a grand amusement park. Augustus Saint-Gaudens consulted on sculpture. Thousands of men, living in barracks, transporting supplies by mule teams, worked long, grueling days to make Olmsted's plan a reality. By 1893, they had dredged and filled more than seven hundred acres of sand dunes and marshes; they had laid pipes and buried electrical lines; and they had constructed four hundred buildings, most of a cheap concoction called staff, composed mainly of jute and plaster of Paris. The architecture was splendid, designed by the country's most prominent architects, including Richard Morris Hunt, Charles McKim, and Louis Sullivan; the Women's Building was assigned to Sophia Hayden, a graduate of MIT. All of the buildings, the planners decided, would be gleamingly white: harmonious and pure. "I don't recall who made the suggestion," chief architect Daniel Burnham said. "It might have been one of those things that occurred to all minds at once, as so often happens."[4] When whitewash was blown onto the staff, the buildings resembled marble; but these structures were not meant to endure, except in the memory of the fair's visitors.

The fair, the planners agreed, was "the third great American event," after the War of Independence and the Civil War, and, like those other two triumphs over adversity, its success depended on cooperation and a patriotic spirit.[5] Certainly the fair aimed to evoke that patriotic spirit in its visitors. Although the architecture imitated Beaux Arts and Italian Renaissance palaces of Europe, the contents of the buildings celebrated the achievements and golden prospects of the New World, technologically, economically, and morally. Since 1492, the fair insisted, America had been moving inexorably toward perfection. With its industrial strength and moral righteousness, the America of the near future, as a *Cosmopolitan* reporter put it, would represent "the ennobling education of humanity—the diffusion of knowledge, the broadening of the sympathies of communities—the better mingling of the country and the town, the elevation of the labors, the expansion of the ambition, the illumination of mind and matter... awakening the dull, inspiring the despondent, cheering the broken, arming the weak for the greatest cause, that of the common good."[6]

The message may seem cloying, but it was echoed by many visitors. The fair's greatness lay not just in its material wonders, but in the spirit of the men and women who traveled days and weeks to arrive at the White City, seeking a glimpse of utopia. For novelist Robert Herrick, the fair offered a respite from fear and worry; it inspired faith. "The people who could dream this vision and make it real," he wrote, "those people from all parts of the land who thronged here day after day—their sturdy wills and strong hearts would rise above failure, would press on to greater victories than this triumph of beauty—victories greater than the world had yet witnessed."[7] William Dean Howells was rapturous after a visit in September. William James heard glowing reports from his friends: "*every one* says one ought to sell all one has and mortgage one's soul to go there, it is esteemed such a revelation of beauty," he told his brother Henry. "People cast away all sin and baseness, burst into tears and grow religious, etc. under the influence!"[8]

America's prowess was most striking in the demonstration of electricity, the harbinger of a new age of boundless technological achievement. While there had been limited evidence of electricity at the Centennial Exhibition in 1876, at the Columbian, one visitor noted, "it is omnipresent... it has a temple of its own, which is filled with the manifold applications of this strange and subtile agent to the arts and conveniences of life."[9] Even outside of its temple, electricity infused the fair. Since Westinghouse had won the commission, his rumbling dynamos, housed in the vast Machinery Building, generated the alternating current that served all public spaces and many of the fair's halls. Electric launches, each seating thirty and powered by batteries, glided through canals and basins; an electric railway transported visitors among exhibit buildings in fifteen trains, each with four open cars carrying one hundred passengers, and attaining, at certain points, thrilling speeds of thirty miles per hour. An electrified moving sidewalk transported visitors along the waterfront; electric elevators transported them within buildings.

At night, the effects of electricity were particularly astonishing: the most powerful searchlights in the world swept across the sky "as if the earth and sky were transformed by the immeasurable wands of colossal magicians."[10] The beams could be seen in Milwaukee, more than eighty miles away. The rim of the giant Ferris wheel, as phenomenal at this fair as the Eiffel Tower had been at the Paris Exposition of 1889, was studded with three thousand blinking lights. The Manufactures Building, covering thirty acres, was lit by arc lamps hung from coronas along the center of the structure, with individual lamps in corridors and galleries. The coronas were massive, hanging 140 feet from the floor and extending up to seventy-five feet in diameter. Throughout Jackson Park, electric fountains, fitted with colored glass filters, offered ethereal displays of water, radiant in a rainbow illumination.

On the evening of May 31, four weeks after the fair officially opened, the Electrical Building was first lit, exceeding anything fairgoers had seen yet. Behind a fifteen-foot statue of Benjamin Franklin at its portal, the building shone as "one blaze of light with myriads of incandescent lamps with revolving wheels displaying all the colors of the rainbow in ever changing hues, and with unseen pens writing mysterious inscriptions on the walls in letters of fire."[11] The Electrical Building, small at just over five and a half acres, contained every marvel of the past fifty years, as well as machinery and appliances yet to be marketed. At the center of the building, the Edison Tower of Light rose seventy-eight feet from a glass pavilion at the base, its colonnade studded with thousands of electric lamps and multicolored globes, and topped with an eight-foot cut-glass incandescent bulb. This florid celebration of Edison, however, did not reflect his domination of the fair; Westinghouse outshone him, not least with an exhibition along the entire south wall of the Electrical Building. The two American electrical giants dwarfed the displays of smaller manufacturers and of French, German, and Italian companies.

By 1893, the telephone was no longer the innovation it had been at the Philadelphia Centennial, although most of the 266,000 phones in America (statistically, four for every thousand people) were used by businesses.[12] Rarely was a telephone found in homes. Nevertheless, it was no longer considered a new technology, and now, in the Electrical Building, its history was duly presented for the edification of visitors. The telephone would soon become obsolete, they learned, replaced by the telautograph, which recorded voice messages on paper—a boon for businesses. But since many visitors had never used a telephone, they took the opportunity to make a call on the fair's special system. One reporter described the experience as magical: a visitor could retreat to "haunted corners where one talks to friends a thousand miles away and enjoys the familiar charm of their voices and the magnetism of their presence."[13] Telephone boxes conveyed the idea that the instrument could be a decorative accessory, designed for Victorian tastes. Its portable model (that is, not attached to the wall) was a "table chest," a decorated oak or cherry box with bronze trimmings and feet. A half-moon-shaped shelf, covered with plush, held the transmitter. The handle rested upon the circuit and the bell was fastened to the box's cover.

Just as the telephone had become an attractive accessory, domestic lighting was designed to harmonize with future purchasers' tastes—and to camouflage the products' mechanical parts. At the Manufactures Building, Louis Comfort Tiffany displayed his art nouveau lamps, their jewel colors softening the glare of the bulbs, and their graceful, organic designs evoking Nature's artistry. Although European visitors to the fair remarked on the sleek, functional designs of American machines, lighting fixtures were elaborate, ornate, and predominantly floral. Intricate bouquets contained bulbs as the stamen of each flower; leaves and tendrils sprayed from light brackets mounted on the wall; chandeliers cascaded from the ceiling, with engraved, colored glass holding bulbs and beaten-copper leaves serving as reflectors; tiny lamps, strung on a vine, twined around pillars, hung in the branches of live trees, or shone softly behind translucent stone in walls. Lamp manufacturers were quick to realize that when domestic lighting became available, it would be purchased by affluent men and chosen by women; catalogues and advertisements promoted the "poetry" of light; magazines published hints on decorating with incandescence to enhance the beauty of rooms, as well as women's complexions.[14]

Telephones and lamps, however, were hardly the greatest draw. Here, too, visitors found the latest models of personal electrotherapy devices—hair brushes, belts, corsets—in an extensive display. Here, for the first time, visitors saw examples of the electrical appliances that they were reading about in magazines, appliances, reporters assured them, that would offer them more control over their environment and certainly more leisure time. The Electric Kitchen featured stoves, hot plates, broilers, and water heaters. Electric washing machines and irons in the laundry (the irons weighing in at a hefty eight pounds), electric carpet sweepers in the parlor—all these inventions would ease housework, even if they required that users become electricity experts. One Wisconsin company displayed a hair curler heated by electricity; another featured a shoe polisher that would do its work while its owner reclined in an easy chair. Throughout the house, electric heat regulated by thermometers meant that in the future residents could calibrate indoor temperature to the exact degree of comfort. As wondrous as these appliances seemed, however, visitors most likely did not question how they would be powered. Without plugs on the appliances, without receptacles on a wall, those broilers and sweepers and hair curlers needed to be wired directly into a house's electrical system. Only after the turn of the century, in those 8 percent of homes that were powered by electricity, would consumers be able to plug an appliance into a ceiling receptacle, from which wires hung down like the tentacles of an octopus.

Still, the fair confirmed a promise: electricity, at last and forever, was being harnessed in the service of humans. "[T]he same mighty, subtle, delicate, formidable agency and mastery permeates the atmosphere that compasses the universe," exclaimed *Cosmopolitan*, "and all this is but one breath of the all-embracing vital air, one sparkle of the surf that is the boundary of oceans, the great deeps beyond, unfathomed, but one may believe not unsearchable."[15]

Searching, though, implies a goal. Despite the fair's motto, "Not Matter, But Mind; Not Things, But Men," the goal that America displayed most blatantly was material progress and heady, if not hedonistic, pleasure. The greatest attraction at the fair was the Midway Plaisance, a gaudy strip of sideshows, curiosities, and entertainments. Its most spectacular feature was the wheel, more than two hundred feet tall, designed by the young bridge builder and engineer George Washington Gale Ferris Jr. Its thirty-six wood-paneled cars, fitted with plush seats, each carried sixty passengers. At fifty cents a ride, the Ferris wheel emerged as the biggest money-maker of the fair, day or night, clear skies or overcast. Although the view was breathtaking, the ride itself was the real attraction. Fairgoers thrilled at turning two revolutions hundreds of feet above the midway, surrounded by sparkling bulbs. Compared to the Ferris wheel, the other exhibitions seemed tame, even the forty-five women in national dress who displayed themselves in what visitors called the "Congress of beauty"; even the eight dancing girls, four sent by the sultan of Solo, four by the Javanese government, who performed seductively in scanty costumes; even the Street in Cairo with its snake charmers and Egyptian tombs; even the Turkish Village with its exotic bazaar and Bedouin camp.

At the Electric Scenic Theatre, lights created the effects "of dawn and sunrise, midday, twilight, moonrise, the night sky gemmed with stars, thunder-storms and fair weather, as seen in the Tyrolean Alps, accompanied by such instrumental music and weird yodeling as the traveler hears in these favorite resorts." Among the midway's curiosities were ethnological exhibitions from nations "civilized, semi-civilized, and barbarous,"[16] which, in the guise of educating the populace, fed racial and ethnic prejudice. It was no accident, those exhibitions taught, that America, white and Protestant, had achieved such greatness in the past four hundred years.[17]

But for most visitors who thronged the fair, that greatness seemed as fragile as the ersatz marble and ephemeral buildings. The fair, scheduled to open on Columbus Day, October 12, 1892, had been delayed for seven months because a severe financial crisis slowed construction: the American economy was slumped in a frightening depression. Banks failed, factories closed, investments declined, incomes plummeted. On the same day that *The New York Times* reported the opening ceremonies, another headline read "STOCKS ON A DOWN GRADE." In the following weeks, stocks slipped even further. For the first two months that the fair was open, it appeared deserted; in the massive Manufactures Building, one visitor observed, there were "barely sufficient people to furnish a congregation for a village church."[18] Even though admission was only 50¢ for adults and 25¢ for children, visitors incurred other expenses. Attractions on the midway charged extra fees: 50¢ each to visit the Javanese Village, Persian Theater, and Bernese Alps Panorama; 25¢ to see South Seas Islands and the scale model of the Eiffel Tower; and more than $1 to enter the sultry Street in Cairo. Add to that the cost of transportation to and around the fairgrounds, and the purchase of food and souvenirs, and a day at the fair became a costly outing, too costly, surely, for the newly unemployed.

The atmosphere of fear and worry that permeated America was intensified in its cities. Beyond the gates of the fair, Chicago bred some of the country's worst slums, where life now became even more precarious. "Chicago asked in 1893 for the first time the question whether the American people knew where they were driving," Henry Adams remarked. But Adams heard no coherent answer. "The Exposition itself defied philosophy... [S]ince Noah's Ark, no such Babel of loose and ill-joined, such vague and ill-defined and unrelated thoughts and half-thoughts and experimental outcries as the Exposition, had ever ruffled the surface of the lakes."[19] Yet if Adams failed to see the fair as a symbol of hope in the future, other visitors sent exuberant reports home to their friends. Gradually, attendance increased, and by the time it closed, twenty-seven million people had visited, half of them Americans. Harvard's professor of art history Charles Eliot Norton was among them; a man not easily pleased, still Norton left inspired: "The Fair, in spite of its amazing incongruities, and its immense 'border' of vulgarities," he wrote to his friend Henry Fuller, "was on the whole a great promise, even a great pledge. It, at least, forbids despair."[20]

The fair, though, did not so much forbid despair as distract visitors from recognizing a pervasive atmosphere of enervation that had spread throughout the nation, an atmosphere that was not dissipated by the fair's exuberant celebration of technology and consumerism. Perhaps, some people thought, fatigue and depression were the causes, and not the results, of the financial troubles of the early 1890s. A month after the fair closed, *Harper's New Monthly* published an editorial noting "a paralysis in the body-politic" and a decided "want of confidence." Despite signs that the country should be prospering, it seemed unable to shake off its "mental infirmity." The only answer, the editors decided, was to enlist the help of mind curers, lay practitioners who had moved into the therapeutic field as heirs to both mesmerism and electrotherapy. Mind curers, imposing onto the patient's mind their own strength of will and power of suggestion, built their reputation on the notion that "if you think you are well, you are well." America needed simply to think itself well. "The mind of the nation is alone responsible for the disasters of the nation... In short, if we stop thinking that anything is wrong, nothing will be wrong." Since mind curers claimed that they could act even at a distance from their patient, the editors implored them to act, and act quickly, on behalf of the whole country, to connect the country, suffering, apparently, from mass neurasthenia, to the "vital currents" of the rest of the world, and to nourish the nation's depleted energy.[21]

IMPROVING THE TIME

Just before the fair opened, the American Press Association asked seventy-four well-known men and women—writers, captains of industry, politicians—to predict the future. Their responses, published in newspapers, repeated several themes.

First, that the future would be characterized by speed. Widespread use of telegraphs and telephones would accelerate communication, trains would whiz along at one hundred miles per hour. Second, labor problems would be resolved—no one suggested how—including the irritation of inadequate and incompetent servants, a recurring complaint. And last, when asked to name the great man of the century, most responders did not hesitate: Why, Thomas Edison, of course.[22]

Yet as much as people admired Edison, as eager as they were to gasp at electrified spectacles, they still worried about the connection between utilitarian and moral empowerment. It was one thing to have a host of inventions, another to eliminate poverty, quash anarchy, calm protestors, and enact true and lasting social reform. Recent economic problems were nothing new in American life. Sporadic economic protests in the 1870s had evolved into a decade rife with dissension: in the 1880s, nearly ten thousand strikes involved more than 700,000 workers. Demonstrations were bitter and often violent. In 1886, Chicago saw the worst of them: the Haymarket Square riot, where a rally for an eight-hour work day turned into a bloody confrontation between strikers and police. Immigrants thronged cities, often living in squalor, where tuberculosis and other infectious diseases spread unchecked. The country's reputation for industriousness seemed to rest on the shoulders of the exploited. Factory workers earning less than $500 a year couldn't afford to buy an electric shoe polisher, nor the richly upholstered chair in which to recline while it shined their shoes; they could barely afford rent and food.

Reformers, reminding Americans of the shared principles upon which the nation had been built, urged cooperation to effect vast and enduring changes. To promote that message of unity, Baptist minister and Christian Socialist Francis Bellamy, an editor of the widely circulated magazine *Youth's Companion*, wrote a salute to the flag that he hoped would be recited on Columbus Day at ceremonies throughout the country. He called it the Pledge of Allegiance: to a nation, indivisible, that promised its citizens, one and all, liberty and justice. Bellamy deliberately omitted the word equality, well aware that his co-editors would never approve publication of the pledge if it hinted that women and African Americans were the equals of white men. Yet Bellamy, like many other Americans, realized that true reform needed to guarantee social and economic equality, and to bestow on every citizen the means to pursue happiness.

Whether equality and happiness could be connected to technological progress was a disturbing question. Would electricity become a factor in social and economic reform, or merely contribute to the exploitation of workers? Would it become a force for moral good, or for evil? In a nation that seemed divisive and fragmented, could electricity serve as a unifying force, empowering rich and poor alike, easing labor, raising the standard of living, and creating social harmony? Even writers fascinated by the potential of electricity imagined the malevolent consequences of the force's misuse.

A year after the nostalgic *Wizard of Oz* appeared, celebrating family and rural America, L. Frank Baum published another fantasy, *The Master Key*. This time the protagonist is a bright and curious boy, Rob, not unlike Baum's own son at the time—or Edison half a century earlier. Rob's adventures begin when he accidentally calls up the Demon of Electricity, who, like any genie, must bestow powerful—in this case, electrical—gifts upon his new master. The Demon assures Rob that electricity is a secret that Nature has not yet revealed, not even to Edison. "His inventions," the Demon says, "are trifling things in comparison with the really wonderful results to be obtained by one who would actually know how to direct the electric powers instead of groping blindly after insignificant effects."[23] The Demon controls unheard-of electrical powers, and he shares some with Rob: electrical tablets for daily nourishment, a special electrical weapon to render an enemy unconscious for an hour, an electrical propulsion device for travel, and a protective garment with the "power to accumulate and exercise electrical repellent force." Enhanced by electricity, Rob leaves for exotic locales where he encounters cannibals, Turks and Tatars, and assorted monsters. What might have been a typical boy's adventure story, however, becomes a vision of a terrifying future. Among the Demon's gifts are an electrical "Record of Events" that can transmit exactly what is occurring anywhere on earth, and a "Character Marker" that reveals the true nature of every person by capturing vibrations sent out by the mind and translating them into a code: W for wise, F for foolish, C for cruel, E for evil.[24] With both the record and marker, privacy is obliterated; anyone's activities can be made public and anyone's mind invaded, all in the service of disseminating truth and knowledge. The Demon even has an "Electro-magnetic Restorer" that frees its wearer from disease and can revive the dead, "provided the blood has not yet chilled."[25]

These new devices, Rob discovers, easily can fall into the wrong hands: an evil scientist wants them, as does a greedy businessman, and neither has the good of humankind as a goal. Despite his initial enthusiasm for the fabulous powers, Rob, with the common sense typical of Baum's young protagonists, decides that civilization is not ready for the Demon's inventions. The Record of Events and Character Marker seem particularly horrifying: "What right have you to capture vibrations that radiate from private and secret actions and discover them to others who have no business to know them?" Rob asks the Demon, repeating, more than fifty years later, the objections that his grandfather might have made to the telegraph. "This would be a fine world," Rob adds, "if every body could peep into every one else's affairs, wouldn't it?"[26] Yet the Demon, though rejected by Rob, predicts that the day will come when he will be released, ready to loose upon the world a technological—and moral—revolution.

In physicist John Trowbridge's novel *The Electrical Boy*, Richard Greatman, born in a "foul tenement," is adopted by an elderly electrician after his mother dies. Richard's earliest memories are electrical: streetlights and the linemen who repair them, and when his mother dies, he thinks she has ascended a light pole to heaven. With the credulity and innocence of a child, Richard sees his guardian as "a magician who employed electricity as the men of the East did the genii in the tales of the Arabian Nights."[27] He comes to learn, however, that the electrician is actually a thief, stealing electricity from a public source and diverting it for his own profit. At first, Richard, too, becomes involved in deceit, making a device that enables men to cheat at cards by telegraphing what others hold; and he discovers that there are other ways that electricity is used for evil ends. When Richard gets a job creating entertainment for the Dime Museum—a magnetic doll, a flying machine, and a melodrama called "Playing with the Devil," that features sword fights with electrical sparks—he sees that an animal tamer who claims to subdue wild animals through the use of magnetism actually is administering painful shocks. Dr. Socrates, an electrotherapist and seer, uses electrical devices to simulate spirits. Nevertheless, Richard and his electrical talents eventually rise above these temptations for misuse, and when he becomes wealthy as an electrician, he uses his riches to aid poor children.

Trowbridge knew better than to suggest that readers reject technological innovations, and even encouraged them to experiment with electricity by including within his tale instructions for assembling batteries and other devices. Yet he cautioned them that there is one force more powerful than electricity, one force that must influence how and why electricity is used: "that force is love."[28]

The force shaping electricity's future, though, seemed just as likely to be greed. Mark Twain was fascinated by electricity in all of its forms. He once wrote an essay testifying to his experiences with mental telegraphy; he gleefully foresaw a day when there would be no books or newspapers, but instead "salacious daily news furnished in a whisper" by phonograph; he visited Tesla's laboratory where he was photographed, a sphere of light glowing between his hands. Eager to capitalize on the public's interest, he considered writing a biography of Edison or a history of the force addressed to "ordinary intelligent people who are not electricians."[29] Instead, Twain offered his readers two satires about the consequences of allowing electricity to fall into the wrong hands.[30] A play, *Colonel Sellers as a Scientist*, written in collaboration with William Dean Howells, and a novel, *The American Claimant*, both feature the wily Colonel Mulberry Sellers, a tinkerer and inventor, interested in nothing more than a quick profit. Among his inventions is a phonograph that curses, "The Sellers Ship's Phonograph for the application of stored Profanity," particularly useful, he boasts, on sailing vessels during storms. As his response to the grave economic problems of the time, Sellers comes up with the idea of materializing the dead to provide cheap labor. If two thousand New York City policemen earn a dollar a day, he will replace them with dead ones at half that cost. Like the fictional Miss Clary, they won't need to eat or drink, "they won't wink for cash at gambling dens and unlicensed rum-holes, they won't spark the scullery maids." Materialization, as Sellers explains it, is based on the theory that every human consists of "long-descended atoms and particles of his ancestors" that can be made to cohere when a current of electricity "takes up unorganized matter whenever it finds it, and clothes the spirit of the departed in it." The process can solve the servant problem, as well: he will kill servants, materialize them, "and after that they will be under better control." It could even improve government. "I will dig up trained statesmen of all ages and all climes," Sellers announces, "and furnish this country with a Congress that knows enough to come out of the rain—a thing that's never happened yet, since the Declaration of Independence, and never will happen till these practically dead people are replaced with the genuine article." Nor will he stop with domestic leaders. All the monarchs of Europe, he says, could well be replaced by the "best brains and the best morals that all the royal sepulchres of all the centuries can furnish."[31] In his ridicule of technology and government, Twain responded to the frustration of his contemporaries about failed leadership to address persistent, endemic problems.

It took a modest journalist to propose a path to reform that seemed so right, so workable, and so revolutionary, that, almost overnight, he became one of the most important writers in the country. Francis Bellamy's cousin Edward Bellamy, the son of a liberal Baptist minister, grew up in Chicopee Falls, Massachusetts, where he saw firsthand the struggles of poor factory workers to support their families. Although his mother hoped that he would follow his father into the ministry, he decided instead to become a lawyer, with the idealistic goal of arguing important constitutional issues, zealously defending the rights of widows and orphans, and vindicating those falsely accused of crimes. But when he finally opened his own office, one of his first cases required him to evict a widow for nonpayment of rent. The prospect was so horrifying that he closed his office and never practiced law again. Instead, he became a writer—of journalism for the *New York Evening Post* and later the *Springfield Union*, and of fiction, earning acclaim for his novel *The Duke of Stockbridge*, based on Shays's Rebellion, in which eighteenth-century farmers in Massachusetts rose up against the greedy bankers, businessmen, and politicians who oppressed them.

In 1886, shaken by the Haymarket Square riot and his own observations of urban poverty, Edward Bellamy was moved to write an even more pointed work, a vision of the country changed beyond recognition, a vision of America as utopia. Married and with two young children—his daughter an infant—he hoped that the world they inherited would be radically different from the world as it was. The many reform movements that vied for popular support seemed to him only to exacerbate divisiveness; what was needed, he believed, was cooperation, true solidarity, full equality, and the overthrow of capitalism. This was the premise that informed *Looking Backward, 2000–1887*.

The novel was a publishing phenomenon. At a time when ten thousand books constituted a best seller, *Looking Backward* sold hundreds of thousands in its first year and millions later in America and, translated, throughout Europe. More startlingly, it inspired the solidarity that Bellamy called for: hundreds of Nationalist Clubs or Bellamy Clubs formed to plan for reform in communities and throughout the country. The Boston Nationalist Club, one of the more influential, counted among its twenty-seven members such respected figures as Thomas Wentworth Higginson, William Dean Howells, and the minister Edward Everett Hale. With Bellamy's help, they shaped a platform that called for government ownership of coal mines, railroads, and utilities, and they exerted their influence on political candidates and voters. For ten years, from the publication of *Looking Backward* until Bellamy's death in 1898, the Nationalists, supported by the Populist Party and feminist groups, energized American politics.

The premise of *Looking Backward* is a familiar tale of time travel. Julian West, suffering from insomnia and, apparently, neurasthenia because of stresses and worries, calls upon the services of Dr. Pillsbury, a "Professor of Animal Magnetism," who puts him under a mesmeric trance. Usually Julian sleeps until his servant, specially trained to reverse the process, awakens him. But in May 1887, his house burns down, and he is interred in a secret subterranean room for 113 years, remaining alive because of the vital magnetic force infused by Dr. Pillsbury. He awakens in 2000 to a city transformed, an urban paradise much like the White City that Bellamy and his readers had yet to see. And the entire country has changed, not only physically but economically and socially. No longer is there private enterprise or competition. The government owns all production, and each worker rotates throughout the "Industrial Army"—so called because its workers are infused with patriotism—where they fulfill jobs suited to their abilities. All labor in the new America is considered dignified and valuable. Child care and housework are provided, so women can take an active role in work and public life. Those who cannot work because of illness or debility receive the same benefits as those who do.

These huge changes have occurred not because of technological advances but because of a consensus of will. Still, like most of his contemporaries, Bellamy was captivated by the novelties emerging from inventors' workshops, and in *Looking Backward*, technology helps to create ease and pleasure. The telephone, for example, connects a sitting room of each home to concert halls, offering music twenty-four hours a day; on Sunday, listeners have the option of hearing a sermon. Debit cards make money unnecessary; huge warehouses provide citizens with all they desire in the way of goods. Electricity generates heat and light. The America of the future is free of chimneys, smoke, and soot. William Dean Howells wished that Bellamy had made "the millennium much simpler, much more independent of modern conveniences, modern facilities. It seemed to me," he wrote, "that in an ideal condition...we should get on without most of these things, which are but... toys to amuse our greed and vacancy."[32] But other readers felt comforted that despite capitalism, material progress would move forward, and if the future meant social reform, they would not have to give up the prospect of toys.

They would, though, have to give up complacency. Julian West claims that without competition, people will feel no motivation to strive and better themselves. Selfishness breeds excellence, and selfishness is inherent in the human spirit. But Bellamy argues against the idea that "greed and self-seeking were all that held mankind together, and that all human associations would fall to pieces if anything were done to blunt the edge of these motives or curb their operation."[33] Instead, he urges readers to free themselves from assumptions about human psychology that kept them from thinking and acting boldly. Shortly after publishing the novel, Bellamy made another foray into science fiction with a short story considering the infiltration of the phonograph into daily life.[34] After falling asleep during a tedious train trip, the narrator "awakens" to discover that at his destination, everyone is carrying a miniaturized phonograph, known as "the indispensable," that records and plays messages and reads aloud the daily newspaper. In his hotel room, he is treated to the sultry voice of a talking clock, fitted with a recording that announces the hour. This device, the narrator discovers, is only one among many efforts to "improve the time... There were religious and sectarian clocks, philosophical clocks, freethinking and infidel clocks, literary and poetical clocks, educational clocks, frivolous and bacchanalian clocks," each emitting a "war of opinion... calculated to unsettle the firmest convictions." Equally unsettling is the use of a phonograph to create automatic pastors, designed to shake hands and to convey "general remarks of regard and esteem" to parishioners. Clergymen, it seems, were busy recording their sermons on phonograph cylinders, too busy, in Bellamy's view, to minister to spiritual needs and provide moral sustenance. Politicians were no better: a phonographic President of the United States greets the public, able to tell visitors exactly what they want to hear about the pressing issues in their home state.

The narrator's dream, innocent and fanciful as it seems, suggests a darker message: the phonograph has "improved the time" by invading privacy, spewing propaganda, and substituting generic patter for considered responses. It has become indispensable by making its users expect the constant stimulation of news, information, and sound. It has made the skills of reading and spelling obsolete; even more ominous, it has become a substitute for face-to-face interaction. People could communicate by recorded cylinder, or with automatons standing in for moral and political leaders.

Common to both of Bellamy's fictions was a call for independent thought and a recognition of the difficulties involved in taking personal responsibility for the common good. Just as Beard had done in 1869, neurologists and electrotherapists in the 1890s insisted that the stresses of living in a democracy were primary factors in depleting energy. After George Beard died, British physicist William Crookes joined other colleagues in praising his contributions to medicine, including Beard's assertion "that America, and to some extent the whole civilized world, in seeking to make every man, woman, and child an expert in politics, is sapping the very vitality of the human race."[35] Men and women would best defer to authority, if only they could find an authority they trusted.

Simon Newcomb, ardent observer of electrical innovations, offered his own fable about responsibility and authority in a novel that appeared in 1900, just after the Spanish-American War, and in the middle of international tensions that seemed to augur a more devastating conflict. Newcomb set his novel in the 1940s, when he imagined, with uncanny prescience, such tensions were likely to have intensified. If humanity needed anything, it was a way to achieve lasting peace. Unlike other writers about the future, Newcomb did not see electricity as the means to endless prosperity, equality, and ease, but as a potential weapon of death and destruction. Newcomb's protagonist, a Harvard professor named Campbell, is certain that he has found a way to prevent that horrifying outcome: he has invented two substances—etherine, which acts on the ether of space in a peculiar new way, and therm, a kind of electricity. These substances, when combined, are able to power amazing machines: a device that enables men to fly through the air and robotic contraptions of "enormous muscular force" that can be manipulated as if they were huge marionettes. The robots look like spindly caricatures of the human body, with projectiles of jointed arms and legs. "Daddy-Long-Legs" is their nickname, and they represent unheard-of new power. In popular parlance, they are known simply as "daddies."

"Daddy" is an endearing and ironic term for a device that can wreak havoc if placed in the wrong hands. The "daddy," after all, cannot protect, or lead, or guide. It can only reflect the intelligence, morality, and wisdom of the humans that move it. Campbell, sufficiently concerned—as were Newcomb's contemporaries—about the lack of such wisdom, refuses to make his two energies available unless the world changes radically: armies must be disbanded and war must end. If humans are not to be subservient to their technological inventions, if they are not to become submissive children to their "daddies," then someone with persuasive moral authority—a strong paternal figure, in fact—needs to take charge. The process is difficult: world leaders, after all, are not eager to relinquish their power and their quest for Campbell's forces, but conflict is relatively short, and after it has died down, Campbell notes, "the weak-minded people had all gone crazy, leaving only those who could keep their heads to look after the world's affairs."[36] At last Campbell assumes leadership: he has styled himself as the Defender. His grateful subjects add an honorific: His Wisdom.

Newcomb voiced concerns that many shared: Who would wield this huge and deadly power? How much would the force change moral values? There seemed a sense of dissonance between what humans could accomplish technologically and what they could accomplish politically, socially, and especially morally. The World's Columbian Exposition had underscored that dissonance, standing as a bright and shining city in the middle of squalor; urging visitors to buy—lights, appliances, and new products like Cracker Jacks, Cream of Wheat, and Juicy Fruit gum—as they feared for their jobs and their savings; intimating a glorious future, yet failing to assuage doubts that such a future could be realized.

The fair had been plagued by distress: financial panic as it was being built, depression during the months it was open, the shocking assassination of Chicago's mayor Carter Harrison just two days before it closed. And more distress followed: in January 1894, the Manufactures Building, Music Hall, Peristyle, and Casino were destroyed by fire. Thc cause was arson; the loss, over $1 million. In July, another blaze rapidly consumed the Administration, Mines and Mining, Machinery, Agricultural, and Electrical buildings. Utopia crumbled to rubble. In the years that followed, the vision of the fair became a dare and a puzzlement: If humankind could achieve this pinnacle of urban planning, why did strikes plague cities? If humankind could invent such technical marvels, why were so many families living in poverty? We sometimes still think the same way: If we can land a man on the moon, why can't we provide better schools at home? If we can map the human genome, why can't we achieve world peace? These connections are not logical, but since the advent of the telegraph, the nineteenth-century public had heard one after another technological innovation touted as the solution to war, dissension, economic downturns, illness, and oppression.

Light, especially, implied moral illumination, a connection expounded most strikingly in an extravaganza called *Excelsior*, which premiered at La Scala in 1881.[37] The performance, according to its choreographer, Luigi Manzotti, was an "Azione Choreographica, storica, allegorica, fantastica": a paean to such thrilling inventions as the steamboat and telegraph, and to such world-changing structures as the Suez Canal; the Mont Cenis tunnel through the Alps, linking Italy and France; and, though still under construction, the Brooklyn Bridge. Among more than a dozen allegorical figures are the prima ballerina, Civilization, and, most centrally, the spirit of Light, who announces to the prince of Darkness that she alone will command the future of the world.

One scene of the ballet takes place in Volta's laboratory, where despite discouragement from Darkness, Volta manages to develop a battery; another scene is set in Telegraph Square in Washington, D.C., where the spirit of Light dances ebulliently, surrounded by scores of telegraph messengers resplendent in lightbulbs that they switch on and off as they leap and bound. The spirit of Darkness lurks in the corners of each scene, awaiting an opportunity to wreak havoc or merely to demoralize inventors; but Light manages to triumph, and in the end engulfs Darkness and all that it represents: ignorance, barbarism, ruthlessness, and death. Light will unite the warring countries of the world, light will bring peace and tolerance, light will bring endless prosperity.

Produced more than a hundred times at La Scala, the ballet traveled to major European cities, where it was offered hundreds of times more. The White City, sparkling, beautiful, and replete with technological wonders, sent a message just as intense. But as the century ended, it became clear that inventions were not the answer to complex problems. Artificial illumination would not necessarily yield enlightenment; for that, many people began to turn inward. Ironically, electricity offered a new way to do so.

CHAPTER 11

Dark Light

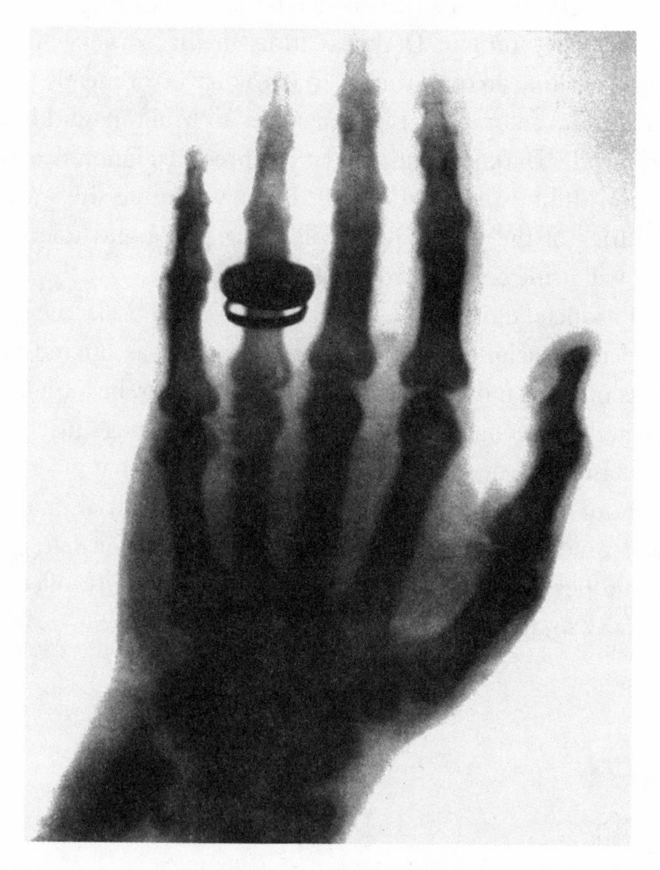

Roentgen's X-ray of the hand of Dr. Albert von Kolliker, professor of physiology at Wurzburg University, January 23, 1896. Roentgen made this image at his first public demonstration, before the Wurzburg Medical Society, generating astonishment and acclaim from his audience.

The Bakken Library, Minneapolis

The question arises, what is this new form of radiant energy?

Science, February 14, 1896

On New Year's Day 1896, Wilhelm Conrad Roentgen, a fifty-year-old professor of theoretical physics and director of the Physical Institute at the University of Wurzburg, sent a photograph of his wife's hand to some physicist friends. He knew his friends would be shocked, they might even think that he had gone crazy; but he was just as certain that they would be thrilled, as he was. What he sent was no ordinary photograph, but a ghostly, ethereal image: the shape of Frau Roentgen's hand appeared as a shadow; yet, astoundingly, the bones were clearly visible, as was the wedding band on her fourth finger. If Roentgen had sent a photograph of his wife naked, it could not have been more sensational. He had discovered a way to penetrate clothing, penetrate the body, and reveal the unseen.

Nearly two months earlier, on November 8, Roentgen had been working alone in his laboratory, investigating cathode rays produced in a Crookes tube, a glass tube fitted with platinum wire at two points. At each point, the wire was connected to a pole of a battery or induction coil: one point (the anode) to the positive pole of the battery, the other (the cathodc) to the negative pole. After air was evacuated from the tube and a very high voltage charged the poles, researchers could see a greenish glow appear on one side of the tube. Others before Roentgen had investigated this phenomenon by placing an object between the cathode and the side of the tube. Because the object left a shadow on the glowing area, it seemed that light, somehow, was involved. This invisible light had become known as a cathode ray. Many researchers at the time, Roentgen among them, believed that investigating cathode rays might lead to understanding the nature of light, and the relationship between light and electricity.

For his investigations, Roentgen had covered the tube with cardboard. Surprisingly, he noticed that a screen coated with a fluorescent material began to glow faintly when electricity charged the poles. The glow was a mystery: neither cathode rays nor visible light could have emanated from the shrouded tube. When Roentgen changed the distance between the tube and the screen, the same glow appeared; even more puzzling, when he held up paper, a thick book, a piece of tinfoil, sheets of rubber, and glass plates containing various amounts of lead, the rays could penetrate some of these apparently opaque materials and, amazingly, the skin of the hand that held them. Roentgen could see a shadow of his bones on the coated screen. Clearly, he concluded, the tube was emitting a new kind of energy. The rays could not be seen or felt; they could not be deflected by a magnet or a prism. What they were, he could not say. He called them, therefore, *die X-Strahlen*, X-rays: the unknown.

By the end of December, he had performed sufficient experiments to write up his findings and submit them for publication in the *Proceedings* of the Physico-Medical Society of Wurzburg. On January 1, he sent the paper to some friends, one of whom leaked the news of Roentgen's discovery to the Vienna *Presse*. Once the London *Standard* picked up the story, it spread rapidly throughout Europe and to America. On January 24, 1896, the journal *Science* summarized reports of the discovery. Roentgen's unassailable credentials included dozens of scholarly papers, but the achievement that he now made public seemed so incredible that *Science* reported it with due caution. "It is claimed," the journal wrote, "that he has found the ultra violet rays from a Crookes' vacuum tube penetrate wood and other organic substance, whereas metal, bones, etc., are opaque to them. It is said that he has thus photographed the bones in the living body, which would be one of the most important advances that has ever been made in surgery."[1] Would be, that is, if true. Although the journal admitted that "any experiments published by him would be accepted without hesitation," the scientific community was eager to confirm Roentgen's claims, and within the week, scientists in Prague, Graz, and London had duplicated his findings.[2]

Even with researchers' excitement mounting, *Scientific American* at first suspected that "there may be less of novelty in the experiments than is generally supposed." But within a month, its editors soon changed their opinion. Roentgen would be immortalized for his discovery, the journal claimed, and "the year 1896 distinguished as the 'Roentgen photography' year... It seemed as if the limits of human discovery were being reached, but the wonder of the new photography only emphasizes the possibility of other victories to be won in the world of science."[3] The Roentgen photography year was characterized by wonder and fear, a reprisal of the public's reaction to the telegraph and electric lighting: amazement at the technology, eagerness to hear vast claims for its significance, popular enthusiasm for its use as entertainment—and resistance to its potential impact on the body and mind.

In America, newspaper and magazine reports about the X-ray focused largely on Edison, sought by the press for his opinion on "the substantial value of the discovery."[4] Unlike some scientists, who thought that X-rays had greater theoretical than practical implications, Edison immediately realized its benefit for the medical community, especially for surgeons needing to locate foreign bodies such as bullets, or for physicians needing to set broken or fractured bones. If there were victories to be won in this new field of investigation, Edison wanted to win them first. Although he was embroiled at the time in mining operations in Pennsylvania, he rushed to his West Orange laboratory as soon as he read about Roentgen's discovery. He wanted to develop his own X-ray machines, he wrote to Arthur Kennelly, "before others get their second wind."[5] Within weeks of Roentgen's announcement, Edison had created a simplified and strengthened Crookes tube, and by mid-March, he had invented a fluoroscope, a version of the X-ray machine that produced sharp images instantly, but did not result in a photograph. Although most physicians preferred X-ray photographs for diagnostic records, Edison's fluoroscope became popular with physicians when a photograph was not needed, and with the public for personal entertainment.

As he and his staff worked intensely on developing a faster and sharper X-ray machine, Edison claimed that he was not motivated by profits. At the end of March, a report about the fluoroscope announced, "The Wizard Worked for Humanity and Will Not Patent His Wonderful Discovery."[6] Instead, he donated his X-ray machines and fluoroscopes to surgeons and physicians, but he also manufactured a hand-held fluoroscope, modeled like a stereoscope, so that people could inspect their bones as a form of amusement. At the National Electrical Exposition, which opened at Grand Central Palace in New York City on May 4, Edison's exhibition gave the public its first glimpse of X-ray technology, along with half-hourly illustrated lectures. Although the lectures aimed at edification, the display itself was designed to intensify awe. In a space about twenty feet square, curtained to create darkness and lit by two red incandescent lights, hundreds of visitors filed through in pairs, placing their hands behind a screen while an operator turned on the current. Each time the X-ray machine was activated a foghorn resounded throughout the exposition; but visitors needed no sound effects to create a sense of the momentous. For about three seconds, they saw the bones of their own bodies made visible by the amazing "new light."

After the exposition, because the technology of Crookes tubes and electrification was widely available, X-ray machines proliferated, and the resulting images—known variously as shadowgraphs, electric shadows, cathodographs, or Roentgen photographs—became the most intimate portrait one could have taken. Friends and lovers took to exchanging them. But to some people, a view of one's skeleton "within the enshrouding flesh," as one observer described it, looked too much like seeing one's bodily remains[7]; these visions of the body's interior seemed analogous to an autopsy, thrilling but ghoulish. Yet the thrill was undeniably tantalizing.

In March 1896, a spectacle, imported form Paris, arrived in New York: "Cabaret du Neant," or Tavern of the Dead. Although the display did not incorporate X-rays, its effects mimicked what people had been reading about for the past few months. Visitors to the eerie cabaret passed through a long hall draped with black to enter a "spectral restaurant" where coffins served as tables, with a burning candle on each end. At the center of the restaurant there hung a chandelier apparently constructed of skulls and bones. As the guests dined, their guide to this underworld directed their attention to pictures lining the walls. When the pictures were lighted, the figures in each appeared as skeletons.

As funereal chimes accompanied their passage out of the restaurant, the guests entered a second chamber, also draped in black, with a coffin standing upright at one end. Each visitor was invited to stand in the coffin, draped in a white sheet. As the spectators watched, the visitor's body dissolved before them, appearing as a skeleton, then transforming back into a corporeal body. In another chamber a visitor was seated at a table, when suddenly a spirit—of an old man or, more enticingly, a seductive woman—appeared at one side. Invisible to the seated guest, the spirit could be seen by the astonished spectators. Other scenes featured ghostly figures who apparently could pass through solid objects. Although the cabaret relied on magic lanterns, Argand burners, and mirrors, it seemed to produce its effects through the use of X-rays and heightened the public's association of X-rays with death.

Yet the press insisted on the X-ray's potential benefits to the living, benefits that had not been proven through laboratory experiments or clinical trials. Speculations became truths merely by repetition. Throughout the summer and fall of 1896, rumors raged about the various and often contradictory powers of X-rays: they could grow hair or act as a depilatory; they could locate and kill cancerous tumors and obliterate bacteria; they could revive the dead; they could give sight to the blind. Edison fueled this idea. One day, when his eyes began to bother him after hours of working with X-rays, he covered his closed lids with his hand and claimed to have seen his bones. Articles reported cases of X-rays restoring the sight of blind men, women, and, touchingly, children, only to follow those reports with staunch refutations by oculists. The blind, they protested, could not see by "X-ray vision," but some could perceive shimmering and ephemeral effects of fluorescence.

Despite these unverified claims for the beneficial uses of X-rays, other articles pointed to dangers. This new "dark light," as some writers referred to X-rays, seemed to be taking science once again to the brink of transgression. If human eyes could not see through opaque matter, perhaps, asserted *The Lancet*, that was nature's "wise provision." Scientists still could not explain the phenomenon of light, and now, the medical journal wrote, "it is obvious that we must attach a deeper and much wider meaning to the word light than has hitherto been ordinarily understood."[8] Although X-ray images were called the new photographs, clearly they were something far different from reproducing perceptible reality. The machine that produced these images, and the photographs themselves, signified to some a frightening new violation of space and privacy. It was one thing willingly to submit to having an X-ray shadowgraph taken; it was another to be subject to furtive penetration. The X-ray machine seemed even more insidious than the camera, popularly known as the "Peeping Tom." Modern life, lamented one magazine writer, "brought people face to face with that prospect of perpetual and unescapable publicity towards which the resources of science and the conditions of modern life have been together tending to hasten them within recent years." Unlike a conventional camera, this device could see through walls into one's home, through clothing into one's pockets, through flesh to reveal one's organs. Worse than the telephone, which could "pursue business men into their libraries and 'ring them up' in the very bosom of their families," worse than the phonograph, which could reproduce "the most intimate of domestic confidences," the X-ray machine, the "crowning menace," could transcend every possible barrier and make it impossible to find "any refuge from the world of outside affairs."[9] With the X-ray, privacy was obliterated. It is no wonder that some New Yorkers campaigned for a law to prevent the use of "X-ray opera glasses," and no wonder that at least one clothing manufacturer offered X-ray-proof underwear. Popular magazines spoofed these concerns in cartoons: fashionable women at a party were depicted as skeletons; servants were shown peering into a home's private spaces; the rich and poor, pared down to their skeletons, were indistinguishable.

If the body could be penetrated so stealthily, it seemed only logical that the mind could, too. The public heard rumors that professors were sending lessons to students through X-rays, the better to imprint them permanently in their minds. Edison announced that he would soon photograph the brain; translated in the press, it seemed that he would make thought visible. The idea was alluring, especially to those seeking proof that brain waves and X-rays were related phenomena. David Starr Jordan, president of Stanford and friend of such psychical researchers as William James, published a piece in *Popular Science Monthly* that drew upon, and satirized, the popular conflation of X-rays and Reichenbach's infamous odyl. The Astral Camera Club, Jordan wrote, had been established to inquire into "man's latent psychical powers" and inherent "odic forces." As soon as the club's members learned about the X-ray from their daily newspapers, they immediately looked for evidence that "the invisible waves sent out through the ether by the mind could also affect a sensitive plate." After all, if thoughts could be transmitted between people and registered in someone's consciousness, if thoughts could be perceived by mediums, then why should those thoughts not affect matter? Evidence was not hard to find. Aided by a special camera called a sympsychograph, members of the Astral Camera Club and other psychical researchers turned their mental energies to producing a variety of homely images—a postage stamp and, unaccountably, the face of a cat. Many people, fearful of X-ray penetration, wanted to believe that the ether could be controlled by the mind and, therefore, that no permutation of electricity could threaten human supremacy in the natural world. "The essence of this story," Jordan made clear to his readers, "lies in its illustration of the power of mind over matter when its forces are concentrated."[10]

Psychical invasion, though, proved not to be the worst effect of X-rays. Within a year of Roentgen's discovery, researchers had devised more than thirty-two different models of X-ray tubes, and written nearly fifty books and more than a thousand papers on their experiments and speculations.[11] Physicians had begun to use X-ray machines to examine broken bones and locate swallowed objects that required surgical removal; some physicians, believing X-rays could kill bacteria, administered lengthy treatments for a variety of illnesses and infections. But by the end of November, disquieting news emerged from some hospitals, laboratories, and universities. Elihu Thomson, certain that experimentation with X-rays would help scientists understand the mysterious "borderland between matter and ether,"[12] was forced to stop his own experiments when, after exposing one of his fingers to an X-ray for half an hour, he had produced swelling, redness, pain, and a large festering blister. Patients who had been subjected to X-rays suddenly found that their hair was falling out. Some patients suffered such severe ulcerations that they needed to have limbs amputated. In Edison's own workshop, one assistant succumbed to radiation burns that eventually caused his death. Many years later, when Edison was an elderly man and a dentist suspected he had abscessed teeth, he insisted on having his teeth extracted rather than undergo diagnostic X-rays.[13] Although in 1896 he claimed that he was developing cheap lighting based on X-ray technology, he gave up that research completely when he saw the destruction it could cause.

Amid articles about the dark light, newspapers readers learned that on October 14, 1896, Franklin Pope, Edison's mentor and former partner, had been electrocuted as he repaired the wiring in his Great Barrington home. If even experts were not safe, ordinary consumers of electricity were that much more vulnerable. Reports of death by accidental electrocution were commonplace by the end of the century, fueling the public's persistent fear of tampering with the mysterious force. Now, though, a new threat emerged: an invisible, imperceptible ray that could deform and destroy the body. Nature, it seemed, had another means of revenge.

LIKELY STORIES

The potential of the X-ray to injure, however, did not keep physicians from using it for a surprising number of ailments, including neurasthenia.[14] X-ray therapy, according to some practitioners, caused the patient to relax, induced sleep, and generally had the same effects as conventional electrotherapy, which, by the end of the century, had become increasingly sophisticated and widespread. Physicians had a profusion of electrotherapeutic instruments at their disposal, more elaborate than earlier battery sets. The French physician Arsène d'Arsonval, for example, invented the solenoid cage, a structure that resembled a huge wire wastebasket, large enough for a patient to stand within it, subjected to a high-frequency oscillating current emanating from a wire helix. The cage was a treatment of choice for diabetes, gout, tuberculosis, and pertussis. Patients exhausted by exertion or the trauma of an operation might find refreshment from treatment on a magnetic couch, which looked like an ordinary chaise longue, except that under the tufted leather upholstery there were heavy horseshoe magnets powered by an electrical current. Electric baths, in which a patient sat immersed in electrified water, were a familiar therapy. Now there were electro-thermo vapor baths, cabinets featuring a pan half filled with heated water, which created steam. On moistened skin, the patient would move electrodes as desired, massaging or stimulating parts of the body. In addition to the new baths, there were electric "showers," where a patient, standing on an insulated platform, was subjected to a "static crown breeze" from a corolla swinging above the patient's head, or a localized static spray, or an interrupted direct current applied to bare skin for a sedative effect or through a wet sponge electrode for a stimulating effect.

Wherever the body provided a means of entry—vagina, urethra, ear—physicians pushed in electrodes. Patients had electrodes sent down their esophagus to treat stomach problems, or inserted into their rectum for intestinal problems. Static, galvanic, faradic; localized or diffused; direct, indirect, surging, or continuous—there was an electric treatment for nearly every complaint. Physicians combined clinical experience with information disseminated through professional channels to help them determine the most effective therapy for their patients. Besides matching a therapy to a particular illness, physicians needed to decide on the frequency and duration of treatments, and to choose adequate strength of the current. To do so, they had to be familiar with the effects of therapeutic applications. "Don't try experiments on patients," one popular medical guide advised. "Try them first on yourself and see how it goes."[15] Even after physicians experimented on themselves, however, they could not know precisely how electricity would affect any particular patient. That response depended in large part on the body's resistance, which varied according to the patient's height and weight, density of skin, bone structure, and gender. Electrical manufacturers offered physicians sensitive measuring devices to ascertain a patient's resistance before prescribing therapy.

Despite careful diagnoses and clinical expertise, physicians found that some patients needed more than treatment by X-rays or electrical infusions. Certainly neurasthenia still prevailed as "the fashionable disease" of the sensitive and intelligent, but Paul Dubois, an influential French physician, suggested that "psychasthenia" would be a more accurate term. He was seeing evidence not of nerve weakness but of failure of will.[16] His patients seemed mired in despair. Bolstering a patient's willpower, according to Dubois, did not depend on infusions of electricity so much as it did on the physician's power to suggest alternative ways of thinking. To that end, some physicians who had relied on electrotherapy began to supplement that treatment with conversations in which their power over their weak, exhausted, indecisive patients came from an exertion of authority rather than an application of electricity. Neurasthenic patients needed to be convinced of their own capacity to change and, importantly, of their own responsibility to do so. "The nervous patient," Dubois insisted, "is on the path to recovery as soon as he has the conviction that he is going to be cured; he is cured on the day when he believes himself to be cured."[17]

Dubois's declaration formed the basis of the many mind-cure therapies, often with spiritual or religious underpinnings, that emerged outside of conventional medicine by the end of the nineteenth century. As William James described mind cure in his *Varieties of Religious Experience*, its dominant feature was "a deliberately optimistic scheme of life" whether that optimism came from a belief in God or one's own self-worth. Its proponents argued for "the conquering efficacy of courage, hope, and trust, and a correlative contempt for doubt, fear, worry, and all nervously precautionary states of mind." As a self-help therapy, mind cure won a large following; James himself had noticed the "Gospel of Relaxation," the "Don't Worry Movement," and "people who repeat to themselves, 'Youth, health, vigor!' when dressing in the morning, as their motto for the day." Those same people forbade themselves to complain about anything, even the weather, and certainly not their negative feelings. Instead, they were convinced that "the powers of the universe" would respond to their own "appeals and needs"; thinking positively could generate positive results. Although James admitted that mind cure never would work for some neurasthenics, others were uplifted; and their success at conquering "the misery-habit" or the "martyr-habit" gave the movement credibility.[18]

Some mind cure therapies, such as Christian Science and New Thought, became so widespread that in 1898 physicians in Massachusetts, in an effort to protect their professional interests, proposed a bill to license all medical practitioners based on examination. Immediately, the public voiced outrage at legislation that they believed would not protect patients from quackery so much as insure physicians' monopoly over healing. James counted himself among the opposition, condemning the bill as "a movement in favor of ignorance" that would "trammel the growth of medical experience and knowledge." Without new ideas from whatever source—ideas such as antiseptic surgery, electrotherapy, or the germ theory of disease, all of which had been controversial—the medical profession would stagnate. James and his family frequently consulted homeopathic practitioners, but he knew that other physicians held such therapists in low esteem. Their opinion, James asserted, was based on ignorance rather than science. "How many of my learned medical friends, who to-day are so freely denouncing mind-cure methods as an abominable superstition," he asked, "have taken the pains to follow up the cases of some mind-curer, one by one, so as to acquaint themselves with the results?" Certainly, he admitted, the mind-cure movement was "religious or quasi-religious." Certainly it depended on the "vital mysteries" of the personal relationship between therapist and patient. But just as surely it achieved results that should not be ignored or trivialized. "The history of medicine," he reminded his colleagues, "is a really hideous history, comparable only with that of priestcraft; ignorance clad in authority, and riding over men's bodies and souls." To rise above that history, physicians needed to show tolerance and to keep their minds open to what they could only imperfectly understand.[19]

DARK LANDSCAPES

As the century closed, what physicians and patients wanted urgently to understand was what James called the hidden self, a self perhaps truer and more authentic than the social self one presented to the world. Yet in this quest for understanding, even those physicians with minds open to scientific discoveries gleaned few insights from the new profession of psychology, focused as it was on experimental rather than clinical research. In newly established psychology laboratories at Johns Hopkins or Clark University, researchers investigated such problems as optical perception and nerve-response time to stimuli. Other questions about the mind—the connection of emotions to physical well-being, and especially phenomena such as dreams, obsessive thoughts, and inexplicable feelings—were being explored elsewhere: by physicians in clinical practice, by psychical researchers, by philosophers who engaged in introspective analysis. Articles about subliminal mental activity were more likely to appear in *The Proceedings of the Society for Psychical Research* than in the *American Journal of Psychology*. The word "subliminal," as James noted, seemed "offensive" to some, "smelling too much of psychical research." Still he and others needed some term to signify that part of the mind different "from the level of full sunlit consciousness." That shadowy realm, according to James, served as "the abode of everything that is latent and the reservoir of everything that passes unrecorded or unobserved." Memories were stored there. All our apparently impulsive, irrational, inexplicable thoughts and behavior, all our dreams, had their sources there. Somehow—and James could not explain the process—the forces contained in the subliminal region could create "tension" that burst into one's consciousness as a "physiological nerve storm" or other "unaccountable invasive alterations."[20] The greatest "vital mystery" of personality, for James, was this dark, swirling subliminal region and its relationship to "sunlit" consciousness.

Because James did not treat patients, his speculations about the subliminal were derived from introspection into his own emotions and psychological "alterations"; from the unconventional psychological research—on automatic writing, for example, or hypnotic trances—that he conducted at Harvard, using students as subjects; and from clinical reports and theoretical papers from physicians in England and Europe. Among them was Sigmund Freud. Practicing as a neuropathologist since 1886, Freud, like James, believed that his patients' problems—physical pains with no apparent organic cause, hysteria, headaches, obsessions—were connected to subliminal memories. "*Hysterics*," he declared emphatically, "*suffer mainly from reminiscences*."[21]

On January 1, 1896, the same day that Roentgen sent out his startling report and X-ray images, Freud sent to his friend Wilhelm Fliess the draft of a paper that he called "The Neuroses of Defense," and subtitled, coyly, "A Christmas Fairy Tale."[22] Eager to propose a groundbreaking psychological theory that would garner him fame and fortune, Freud believed he had derived one. From his treatment of patients suffering from hysteria, obsessional neurosis, and paranoia, he was ready to argue that these afflictions were aberrations of normal psychological states incited by disturbing memories. Like James, he did not understand the process by which those memories became transformed into physical symptoms, but had suggested, in an earlier work, that electrical force was implicated. While the healthy mind is able to defend itself against such memories and limit their effects, some memories seem so insidious that they cannot be controlled. The events remembered, which Freud believed occurred before puberty and were sexual in nature, are able "to release fresh unpleasure." The only defense against them is repression. And repression transforms otherwise transient feelings of guilt, self-hatred, grief, fear, or shame into physical symptoms, compulsive behavior, or a variety of phobias.

By the time Freud constructed this theory, he had had years of experience in treating these patients. Although he used electrotherapy early in his practice, along with hydrotherapy and massage that he administered himself, he came to believe that these physical treatments achieved only short-term results because they did not address the real cause of the disturbance. Much more effective was a therapy that he called the cathartic or pressure technique, a technique analogous to the use of an X-ray to investigate the body: it gave the physician access to the dark landscape of the subliminal. If his patients could be forced to bring offending reminiscences into consciousness, to give words to a visual image, to describe the memory as a story, then the memory would lose its malignancy. This cathartic technique, which he called psychotherapy and which one patient called the "talking cure" or "chimney-sweeping," releases "*the operative force of the idea*" that had been festering subliminally "*by allowing its strangulated affect to find a way out through speech.*"[23]

At first he thought the cathartic therapy could be effective only if the patient was hypnotized; but because some patients would not fall into a hypnotic trance, clearly another approach was needed. Here Freud developed his pressure technique: as he pressed his hands against the patient's forehead, he insisted that the patient relate, freely and spontaneously, "pictures and ideas." If the patient claimed that nothing occurred to her, Freud took it to mean that she was suppressing or censoring an idea, concealing it, or, worse, judging it to be irrelevant. But Freud kept pressing and insisting: only he could judge the significance of the patient's thoughts. Sometimes his insistence worked; at other times, the patient simply refused to yield, and the session became a power struggle between the resistant patient and the aggressive physician. It was a struggle that Freud was not prepared to lose.

Resistance was as significant for the psychotherapist as it was for the electrotherapist: evidence of the boundary between healer and sufferer, an obstacle to be overcome before the physician could overwhelm the patient with his power. If the electrotherapist had instruments to measure a patient's resistance before beginning therapy, the psychotherapist needed to be alert to resistance at every step in the process. As the resistant patient fought to maintain autonomy and privacy, the physician pressed forward, physically and psychologically, with what Freud called an "attack."

The contest of wills took place in a perilous, labyrinthian site, where the intrepid physician and the reluctant patient, guided by the tenuous threads of the patient's narrative, descended into the past. Sometimes the thread brought them to "a stratum which is for the time being still impenetrable. We drop it," Freud said, "and take up another thread, which we may perhaps follow equally far." The threads, as threads do, inevitably became so gnarled and tangled that they created a block, which somehow had to be "melted" or otherwise destroyed. The task was daunting. "It is easy to imagine how complicated a world of this kind can become," Freud admitted. "We force our way into the internal strata, overcoming resistances all the time." Digging and advancing, backtracking and discovering "side-paths," eventually the physician reached a point where he no longer needed to descend through strata, but finally could "penetrate by a main path straight to the nucleus of the pathogenic organization. With this the struggle is won, though not yet ended." The tangled threads still had to be separated; each must be examined. But at this stage, Freud proclaimed the physician victorious, and the patient, finally overcome, "helps us energetically. His resistance is for the most part broken."[24]

Patients had many reasons to resist: some distrusted a process with which they had no experience and a physician whose method of healing was decidedly unconventional. Some patients shared their contemporaries' belief that an illness is real only if it is organic, and Freud's therapy was based firmly on the premise that distress was incited by the affliction of memory and, perhaps, the patient's forbidden desires. Resistance was fueled, too, by the patient's conviction that his feelings and pain were so idiosyncratic that words failed to convey them. Transforming feelings into words seemed to diminish or trivialize them. Words could be misinterpreted. In any case, psychic pain isolated these men and women, and their inability to communicate about that pain isolated them further. A neurasthenic, Freud noticed, "is clearly of opinion that language is too poor to find words for his sensations and that those sensations are something unique and previously unknown, of which it would be quite impossible to give an exhaustive description."[25]

This focus on uniqueness, which Beard and Rockwell had noted years earlier, more than a symptom only of neurasthenia, seemed to be a symptom of the age: protest against a universe composed of atoms, chemicals, and physical processes. "The spirit and principles of science are mere affairs of method," William James wrote, hopefully; "there is nothing in them that need hinder science from dealing successfully with a world in which personal forces are the starting-point of new effects."[26] Such a world would contain endless sources of mystery.

ENIGMA

Yet this celebration of personal forces made some people uneasy. By the end of the nineteenth century, egoism had become a contested term, for both psychology and culture as a whole.

Psychotherapy demanded and affirmed a trenchant self-consciousness. Some people found this prospect irresistible; others were in such emotional or physical pain that—like Harriet Martineau when she resorted to mesmerism—they would try anything for relief. But others were uncertain that listening to one's body and probing into repressed memories were productive and responsible activities. Roberts Bartholow had raised this issue years earlier; now, considering Freud's "pressure technique," James Jackson Putnam, electrician and neurologist at the Massachusetts General Hospital, suggested that psychotherapy "as a basis for treatment... could easily be overestimated." While he accepted the premise that insight into a patient's past could help to explain present behavior, he was not convinced that "special experiences, no matter how important, need to be specifically reckoned with." More important, he was suspicious of any therapy that rendered the patient so dependent on the physician—"an evil," as he put it—which left open the possibility that the physician could implant ideas about the "sexual origin" of distress in the patient's mind.[27] Therapy, Putnam believed, should create self-reliance. Self-consciousness, after all, could spiral dangerously into self-absorption, making individuals uninterested in contributing to the common good and only in illuminating the vagaries of their own mind. No matter how painful one's thoughts, how haunting one's dreams, how bitter one's self-hatred, the person who could function productively as a responsible and moral member of the community was, in society's eyes, mentally healthy.

According to those suspicious of self-analysis, egoists lived for their own pleasure, attentive only to their needs and desires. Some psychological theorists attributed this apparently infantile form of behavior to degeneracy, a throwback to an earlier evolutionary stage.[28] When degeneracy was inherited, little could be done to ameliorate its effects, and some therapists, Putnam and Freud among them, rejected the diagnosis, believing it fatalistic and deterministic. But even they admitted that the egoism of degeneracy, like neurasthenia, could be induced by the stresses of industrial life: the dazzle, the clamor, the incessant news and incessant work, the complexities of social problems—all these could generate a yearning to retreat, to divert one's gaze inward. If the world seemed incomprehensible, one could try to understand one's self; if the world seemed uncontrollable, one could try to control one's own thoughts and feelings. Although neurasthenics frequently included feelings of isolation and loneliness in their roster of symptoms, for others, it was not a complaint but a desire. Withdrawal and detachment became goals for those seduced by the possibility of exploring hidden layers of personality.

Men and women who engaged in self-exploration, like those who were drawn to mesmerism, often saw themselves as especially sensitive, even artistic and creative. It is no surprise, then, that journeys inward resulted in art, literature, and music celebrating the creator's idiosyncratic perceptions, or that some of these expressions were criticized as self-indulgent and even unintelligible. Self-analysis seemed to spawn an aesthetics of degeneration, of which August Strindberg's *By the Open Sea* was exemplary.

Published in 1890, the novel offered a case history of a consummate neurasthenic in its protagonist Alex Borg, superintendent of fisheries, a man of acute sensitivity: self-conscious, self-analytical, and despondent. Like many others "born and bred in the epoch of steam and electricity," Borg suffered from "nervous affections, a fact which was not strange when one considers that he was called upon to destroy millions of antiquated impressions stored in the brain-cells, carefully examine and reject a number of old fallacies, whenever he was faced with the necessity of forming a fresh opinion."[29] He is faced with this challenge frequently, since the world that he encounters daily is not the world for which his genteel upbringing has prepared him. It is a world that has changed economically, socially, and spiritually. Striving for professional success in a time of unbridled competition is one cause of his stress; defining his manhood, socially and sexually, is another; religion is a third. Borg could imagine "the sensitiveness of the crab changing its shell, or the bird when it is moulting": like them, he needs to slough off the armor of received wisdom and antiquated rules, and to re-create himself as a new man.[30] It proves to be a painful process. To defend himself as much as possible, he makes a great effort to keep his appetites—for food, for sex—under strict control, and he shuns human contact. The touch, even the smell, of other people disgusts him. The only thing that manages to soothe his hypersensitivity is intellectual work, not surprisingly in the natural sciences, where he has devoted himself to collecting and categorizing specimens of birds and fish; he plans someday to apply his skills to studying humans—by collecting photographs and soliciting from undertakers, clothing makers, and hat manufacturers information about the measurements of skulls and bodies. With this cache of data, he believes that he can make some scientific conclusions about the human race. But science, however much Borg believes in its potential to yield knowledge, cannot generate the emotional peacefulness that he seeks.

Before he can embark on this project, Borg is assigned to a remote northern fishing village where he must persuade fishermen to give up trying to catch stromming, which have been depleted, and turn their efforts to salmon. His advent into the community is met with hostility, and Borg, although he professes to dislike human contact, is unprepared for the fishermen's rejection and overwhelmed by "[a] feeling as if a thunder cloud of opposite electricity were hanging over his head, disturbing his nervous fluid, trying to destroy it by rendering it ineffectual." Unable to find any sympathy in the village, he is relieved when two vacationers arrive for a short stay. Like many middle- and upper-class Scandinavians, this mother and daughter seek the simplicity of rural life to restore their "worn-out clockwork." In the fishing village they can associate with people who pose no competition for their place in society, but rather serve to remind them "daily and hourly of their hard-earned point of vantage," and reassure them that they have risen far above the lower social and economic classes.[31] The fishing village, the two women believe, will provide a pleasant respite from the tensions of urban life.

The daughter, Marie, needs such a respite desperately, suffering as she does, Borg discovers, from "that soul-sickness which...bears the vague name of hysteria." Her "hysteria," compared with his own soul sickness, seems trivial, even contemptible. "A little pressure on the will; a wish not fulfilled; a plan crossed; and immediately there follows a general malaise," he thinks, "during which the soul attempts to find the cause of the pain in the body, without being able to localize it."[32] He has devised a cure for such cases that he thinks far more effective than hypnosis or change of diet: a dose of the bitter resin asafetida, which produces a profound physical repulsion. As Marie struggles to eject the drug from her body, she will "forget her imaginary ailment" and congratulate herself on overcoming the horror of the treatment.[33]

Borg's malaise, the reader learns, has been generated not merely by crossed plans or unfulfilled wishes, and not merely by the intensity of a world powered by machines and electricity.

Borg has lost faith in God—and without God, Borg saw a universe reduced to the material and mechanical. "Do you know what God is?" he asks desperately of a priest. "He is the fixed point for which Archimedes longed, on which to build up the earth. He is the magnet in the earth which we postulate, because without it the movements of the magnetic needle would be inexplicable. He is the ether which we had to invent to fill empty space. He is the molecule without which the chemical laws would be a miracle. Give me more hypotheses," he begs, "above everything else give me the fixed point outside myself, for I am cut quite adrift."[34]

Morbid exhaustion, egoism, lost faith, all the symptoms of neurasthenia or, if one believed in it, degeneracy were responses to a world transforming in frightening and disorienting ways. When sixty-year-old Henry Adams visited the Paris Exposition in 1900, he found the technology on display vastly changed from what he had seen at the Columbian Exposition. "In these seven years," he said, "man had translated himself into a new universe which had no common scale of measurement with the old. He had entered a supersensual world, in which he could measure nothing except by chance collisions of movements imperceptible to his senses, perhaps even imperceptible to his instruments, but perceptible to each other." A revolution was occurring, Adams decided, far greater than that incited by Copernicus or Galileo or Columbus. This revolution came from inquiry into the ether and the discovery of atomic particles. In medieval times, forces of energy such as magnetism and electricity were thought to be "immediate modes of the divine substance." Now, if scientists explained these "occult, supersensual, irrational" forces as matter, the divine might be lost forever.[35]

Even Rudyard Kipling, a fan of material progress, reflected

his culture's fascination with the inexplicable in his short story "Wireless." One dark and wintry night, a certain Mr. Cashell set up instruments to send and receive messages by Hertzian waves. He understands nothing about wireless technology; in fact, he tells a visitor, no one understands anything about electricity. But he does know that his instruments "reveal to us the Powers—whatever the Powers may be—at work—through space—a long distance away." As Cashell waits for his instruments to relay distant communications, his co-worker Mr. Shaynor, a consumptive who has fallen asleep after drinking chloric ether, suddenly begins to utter and write communications of his own: lines of poetry. Keats's poetry. When Shaynor awakens, he is unaware of what he has written and, he says, is unfamiliar with Keats's work. The visitor, naturally, is mystified by this apparent channeling of a dead poet, and his wonder deepens when he discovers that Cashell, who has been frustrated in his efforts to receive the messages he expected, finds that his instruments somehow have picked up communications between two ships off the Isle of Wight. The two occurrences are equally inexplicable. Who knows why Shaynor has become a medium for Keats; who knows why a wireless set in an apothecary shop intercepted ships at sea. Who knows what messages are swirling in the vast, dark universe. "God knows," Cashell says, "—and Science will know tomorrow."[36]

That promise contained a threat. Electricity, gravity, magnetism, nerve force—all these energies remained as mysterious at the end of the nineteenth century as they had been for Newton. But these mysteries were in themselves illuminating. Paradoxically, they implied meaning; they inspired faith: that an incorporeal Intelligence had created coherence; that some plan, some unity connected apparently disparate phenomena; and, most important, that nature could guard its secrets.

In November 1896, after nearly a year of exuberant articles about the X-ray, the *North American Review* published "The Animal As a Machine," a summing up of current knowledge about the living body written by Robert Thurston, a Cornell professor.[37] Despite advances in medical research and technology, including X-ray photographs of organs and bones, the "vital system," Thurston wrote, "remains to-day the most mysterious of all the wonders of creation." The living organism was the most complex and intricate machine, "self-perpetuating; self-repairing; capable of performing tasks of the utmost difficulty," serving as "a vehicle for the contained and directed soul." Despite years of anatomical and biological research, scientists still had "no positive clue to the nature of that mysterious force which flexes the muscles, bends a finger, moves a limb, . . . still less of the method of telegraphy which directs it" or of the complicated intellectual forces "hidden in the most inaccessible recesses." Modern medicine knew little more about the vital force than the ancient Greeks, Thurston maintained; the living being "remains still one of the most attractive, tantalizing, and important of all the unsolved mysteries of science."

Early in the new century, a German inventor displayed a six-foot-tall "artificial man" constructed of 305 compartments, 7 motors, and 45 accumulators that energized it with 84 volts. Its internal mechanism, modeled on the semicircular canals of the ear, kept the automaton balanced. Although it could neither speak nor respond to sound, it could walk, ride a bicycle, and, as its "star feat," it could write its name: Enigmarelle. "But it is misnamed—there is no enigma in it," reported one popular magazine. "It is nothing but an ingenious piece of mechanism."[38] Throughout the nineteenth century, as one after another ingenious mechanism emerged into Western culture, as artificial illumination and X-rays threatened to obliterate every dark corner of the spirit and thrust every individual into unremitting glare, humans held fast to a belief in the enigmatic essence of nature.

As some inventions and innovations have become accepted, familiar, and even indispensable, still newer technologies continue to generate apprehension. Neurasthenia is no longer an accepted diagnostic term in Western medicine, but depression and various anxiety disorders persist; a rendition of their symptoms would sound identical to what George Beard heard from his patients. Like our nineteenth-century forebears, we feel, at times, overwhelmed, powerless, and exhausted. In part those feelings stem from the media's thrilled or ominous—but always hyperbolic—reporting about the new. In part our concerns are generated by confusion about whom to trust when experts disagree, sometimes violently, about how technologies might affect us, our children, the world. Some of us assuage our anxiety by cherishing a belief in those "wild facts" in the universe that we hope will keep science humble. Unsolvable enigmas console us: sentience and consciousness, growth and movement, and, most especially, the enduring wonder of life itself.

Appreciation

My thanks to the American Philosophical Society for a generous sabbatical fellowship that allowed me a full year to complete this book; to the Bakken Library and Museum for a Visiting Research Fellowship and access to their incomparable collections on electricity and life; to the New York University Faculty Resource Network for the opportunity to be a scholar in residence; and to Skidmore College for both a sabbatical leave and a Faculty Development Grant. For their warm and kind support of this project, I am grateful to Arthur Kleinman, Presley Professor of Medical Anthropology at Harvard University; and John J. McDermott, University Distinguished Professor of Philosophy at Texas A & M University.

My colleagues at Skidmore College were enthusiastic about this project from the start, and I feel privileged to be part of Skidmore's energizing and nurturing intellectual community. Special thanks in the English Department to Terence Diggory, Robert Boyers, and Sarah Goodwin for their efforts on my behalf and for their faith in me and my work, and to Mason Stokes for his kind willingness to read several chapters and his characteristically sharp and insightful responses; in the Physics Department, to Mary Crone, for indispensable help about matters electrical, physical, and atomic; in the Dean's Office, to Charles

Joseph, Dean of Faculty and Vice President for Academic Affairs, and Sue Bender, former Associate Dean of Faculty, for saying yes along the way; at the library, to Amy Syrell, for patiently facilitating interlibrary loans; and at the Tang Museum, to Ian Berry, for inviting me to share my work in a Tang Dialogue with artist Jeanne Silverthorne.

Friends and colleagues helped by suggesting sources: thanks to Barbara Black, Jordana Dym, Penny Jolly, Mary Lynn, Jay Rogoff, Lanier Smythe, Laura Taxel, Alan Wheelock, and Adrienne Zuerner.

I conducted valuable research at the Bakken Library with the help of David Rhees, executive director; Ellen Kuhfield, curator of instruments; and Elizabeth Ihrig, librarian. Their expertise and knowledge of the collections made my visit extraordinarily productive. At the Schenectady Museum, archivist Laura Lee Linder and her staff were welcoming and helpful, as were Danelle Moon and the staff of the Manuscripts and Archives Division of Yale University's Sterling Library, who facilitated my access to the papers of George Beard.

My editor, Andrea Schulz, brought to her reading of this book clarity, sharpness, and perceptiveness that was nothing less than dazzling. It was a pleasure to work with her, and, as ever, with my agent Elaine Markson. My thanks to both for helping me turn this idea into a reality.

My husband, Thilo Ullmann, has inspired me by his astonishing breadth of historical knowledge; he has been a trusted reader at every stage. My son, Aaron Simon, always makes me feel that I am doing something exciting and important: for his boundless enthusiasm, I am both amazed and thankful. Thanks as well to my father, Samuel Perlin, who has awaited this book with interest that I hope will be rewarded.

Endnotes

Please note that the Edison papers are available online at http://edison.rutgers.edu.

INTRODUCTION

1. *Building a Resident Lighting Business*. New York: Society for Electrical Development, 1923.

2. By "the public," I mean those who might have been Edison's domestic customers and George Beard's patients, people who were literate enough to read the newspapers and periodicals that I cite as sources; people who could be characterized—to use Beard's term—as "brain workers," because, professionally or at leisure, they had a connection to the printed word. Those who read the documents that I have used to re-create this period are, for the purposes of this book, "the public," since they derived part of their understanding of their own time from those documents.

3. Isaac Newton, *Opticks*, 4th ed. (New York: McGraw-Hill, 1931 [1730]): 354.

4. Galvani quoted in Marcello Pera, *The Ambiguous Frog: The Galvanism-Volta Controversy on Animal Electricity*, trans. Jonathan Mandelbaum (Princeton: Princeton University Press, 1992): 81.

5. Volta quoted in *Dictionary of Scientific Biography*, vol. 14 (New York: Scribner's, 1976): 79.

6. Coleridge quoted in Kathleen Coburn, *Experience into Thought: Perspectives in the Coleridge Notebooks* (Toronto: University of Toronto Press, 1979): 82.

7. Ibid.: 50.

8. Medwin quoted in Nigel Leask, "Shelley's 'Magnetic Ladies': Romantic Mesmerism and the Politics of the Body," in *Beyond Romanticism: New Approaches to Texts and Contexts, 1780–1832*, eds. Stephen Copley and John Whale (London: Routledge, 1992): 66–67.

9. See Michael R. Lynn, "Enlightenment in the Public Sphere: The Musée de Monsieur and Scientific Culture in Late Eighteenth-Century Paris," *Eighteenth-Century Studies* 32, no. 4 (1999): 463–76.

10. Joseph Priestley, *The History and Present State of Electricity*, 3d ed., vol. 2 (New York: Johnson Reprint Corporation, 1966 [1775]): 134.

11. Davy quoted in David Knight, *Humphry Davy: Science & Power* (Oxford: Blackwell, 1992): 24–25.

12. Coleridge to Robert Southey, 11 August 1801, *Collected Letters of Samuel Taylor Coleridge*, vol. II, ed. Earl Leslie Griggs (Oxford: Clarendon, 1956): 751.

PART I: WONDERS

CHAPTER 1: WORKING GREAT MISCHIEF

1. Samuel Morse to his parents, 2 May 1814, *Samuel F. B. Morse, His Letters and Journals*, vol. I, ed. Edward Lind Morse (Boston: Houghton Mifflin, 1914): 132–33.

2. Ibid.: 133.

3. *Samuel F. B. Morse*, vol. II: 47.

4. Ibid., 27 August 1837: 39.

5. Ibid., Morse to F. O. J. Smith, 15 February 1838: 84.

6. Ibid.: 195.

7. James Wynne, "Samuel F. B. Morse," *Harper's New Monthly Magazine* 24, no. 140 (January 1862): 224.

8. Morse quoted in "Telegraphs and Progress—The Cause," *The Economist* (18 October 1857); reprinted in *Littell's Living Age* 52, no. 658 (3 January 1857): 59.

9. Henry James, "In the Cage," in *Complete Stories, 1892–1898* (New York: Library of America, 1996): 866.

10. Cooke quoted in Iwan Rhys Morus, "The Electric Ariel: Telegraphy and Commercial Culture in Early Victorian England," *Victorian Studies* 39 (1996): 350.

11. James: 847.

12. Reprinted in *Littell's Living Age* (4 January 1845): 21.

13. Frank L. Dyer and Thomas C. Martin, *Edison: His Life and Inventions*, vol. I (New York: Harper & Bros., 1910): 8.

14. See Menaham Blondeim, "When Bad Things Happen to Good Technologies: Three Phases in the Diffusion and Perception of American Telegraphy," in *Technology, Pessimism, and Postmodernism*, eds. Yaron Ezrahi et al. (Amherst: University of Massachusetts Press, 1994): 82.

15. Quoted in review of Ezra S. Gannett's *The Atlantic Telegraph: a Discourse delivered in the First Church, August 8, 1858*, "The Oceanic Telegraph," *North American Review* 87 (October 1858): 544.

16. "The Magnetic Telegraph," *Albany Argus*; reprinted in *Littell's Living Age* 3 (4 January 1845): 34.

17. Ralph Waldo Emerson, *Journals of Ralph Waldo Emerson 1866–82*, vol. XVI, eds. R. A. Bosco and Glen M. Johnson (Cambridge: Belknap Press of Harvard University Press, 1982): 45.

18. "Influence of the Telegraph Upon Literature," *The U.S. Democratic Review* 22 (May 1848): 119.

19. See for example, "The Magnetic Telegraph"; "The Magnetic Telegraph—Some of its results," *Tribune*; reprinted in *Littell's Living Age* 6 (26 July 1845): 63; and "The Telegraph Between New York and London," *International Magazine of Literature, Art, and Science* 1 (1 November 1850): 04A.

20. "The Magnetic Telegraph": 63.

21. Nathaniel Hawthorne, *The House of the Seven Gables* (New York: Penguin, 1986 [1851]): 264–65.

22. "Contributor's Club," *The Atlantic Monthly* 41, no. 246 (April 1878): 544.

23. Ralph Waldo Emerson, *Journals of Ralph Waldo Emerson, 1847–1848*, vol. X, ed. Merton M. Sealts Jr. (Cambridge: Belknap Press of Harvard University Press, 1973): 38.

24. "An Evening with the Telegraph-Wires," *The Atlantic Monthly* 2, no. 11 (September 1858), 489–95.

25. "The Oceanic Telegraph," *North American Review* 87 (October 1858): 534. In *Networking: Communicating with Bodies and Machines in the Nineteenth Century* (Ann Arbor: University of Michigan Press, 2001), Laura Otis writes that "scientists' electrophysical understanding of the nervous system closely paralleled technological knowledge that allowed for the construction of telegraph networks" (12) and suggests that some people "may genuinely have seen animal magnetism and electromagnetism as identical" (129). Statements such as the quoted passages in *North American Review* support a stronger argument: rather than being parallel or metaphorical, language used to describe nervous action and the telegraph reflect popular perception that animal electricity and artificial electricity were identical.

26. W. D., "Touching the Lightning Genius of the Age," *American Whig Review* 14, no. 81 (September 1851): 227–35.

27. Ralph Waldo Emerson, *Journals of Ralph Waldo Emerson, 1843–1847*, vol. IX, eds. R. Orth and A. R. Ferguson (Cambridge: Belknap Press of Harvard University Press, 1971): 260.

28. "Social Electrical Nerves," *Nature* (28 February 1878): 306.

29. Thomas Edison, *The Diary and Sundry Observations of Thomas Alva Edison*, ed. Dagobert D. Runes (New York: Philosophical Library, 1948): 216.

CHAPTER 2: BENEFICENCE

1. Henry Lake, "Is Electricity Life?," *Popular Science Monthly* (1873): 478.

2. Thomas Edison, *The Papers of Thomas Alva Edison: The Making of an Inventor, February 1847–June 1873*, vol. I, eds. Reese V. Jenkins et al. (Baltimore: Johns Hopkins University Press, 1989): 629–30.

3. George Parsons Lathrop, "Talks with Edison," *Harper's* (February 1890): 429.

4. Edison: 634, 636.

5. Ibid., letter attributed to Edison, 16 October 1869: 139, 637.
6. Ibid.: 639.
7. Ibid.: 644.
8. Edison, vol. II: 789.
9. George Bernard Shaw, *The Irrational Knot* (New York: Brentano's, 1918): ix–xi.
10. Robert Bruce, *Bell, Alexander Graham Bell and the Conquest of Solitude* (Boston: Little, Brown, 1973): 36–37.
11. Sir William Thomson, quoted in Bruce: 198.
12. Edison, vol. III: 694.
13. Ibid.: 441.
14. *The Manufacturer and Builder* 9, no. 2 (December 1877): 268; 10, no. 8 (August 1878): 182; and 12, no. 2 (February 1880): 80.
15. Edison, vol. III: 699.
16. "The Talking Phonograph," *Scientific American* (22 December 1877); reprinted in Edison, vol. III: 670.
17. Edison quoted in Frank L. Dyer and Thomas C. Martin, *Edison: His Life and Inventions*, vol. I (New York: Harper & Bros., 1910): 209.
18. Thomas Edison, "The Phonograph and Its Future," *North American Review* 126 (May–June 1878): 527–36.
19. "Contributor's Club," *The Atlantic Monthly* 41, no. 246 (April 1878): 543.
20. Dyer and Martin: 213.

CHAPTER 3: WILDERNESS OF WIRES

1. Thomas Edison, *The Papers of Thomas Alva Edison: The Making of an Inventor, February 1847–June 1873*, vol. IV, eds. Reese V. Jenkins et al. (Baltimore: Johns Hopkins University Press, 1989): 456–57. See also *The New York Mail* (10 September 1879) at Thomas Edison Papers Web site, http://edison.rutgers.edu
2. Cologne *Zeitung* quoted in Matthew Luckiesh, *Artificial Light: Its Influence Upon Civilization* (New York: Century, 1920): 158.
3. Robert Louis Stevenson, "A Plea for Gas Lamps," in *Travels and Essays*, vol. 13 (New York: Scribner's, 1917): 165–69.
4. Quoted in Ido Yavetz, "A Victorian Thunderstorm: Lightning Protection and Technological Pessimism in the Nineteenth Century," in *Technology, Pessimism, and Postmodernism*, eds. Yaron Ezrahi et al. (Amherst: University of Massachusetts Press, 1994): 68.
5. Charles F. Brush, "Some Reminiscences of Early Electric Lighting," Hammond Historical File, Schenectedy Museum, 18 April 1928.
6. Ibid.: 5, 6.
7. "Edison's Newest Marvel," *The Sun* (16 September 1878); reprinted in Edison, vol. IV: 503–5.

8. J. P. Morgan to Walter Burns, 20 October 1878, quoted in Jean Strouse, *Morgan: American Financier* (New York: Random House, 1999): 181.

9. Robert Friedel and Paul Israel, *Edison's Electric Light: Biography of an Invention* (New Brunswick: Rutgers University Press, 1986): 22.

10. "Prof. Farmer's Views on Electric Light vs. Gas," *The Manufacturer and Builder* 11, no. 1 (January 1879): 22.

11. "Gas Versus Electricity," *Nature* (23 January 1879): 261–62.

12. Brush: 8.

13. Ibid.: 10.

14. Peter Tocco, "The Night They Turned the Lights On in Wabash," *Indiana Magazine of History* 95, no. 4 (December 1999): 350–63.

15. "Lights for a Great City," *The New York Times*, 21 December 1880: 2.

16. "Darkness for Half an Hour," *The New York Times*, 24 August 1881: 8.

17. "Edison Electric Light Company," historical document, Schenectady Museum.

18. "Has Edison Really Accomplished Anything?," *The Sun*; reprinted in *The Manufacturer and Builder* 11, no. 9 (September 1879): 207. Similar doubts were expressed in "Gas Versus Electricity," *Nature* (5 December 1878): 107; *Nature* (23 January 1879): 261–62; and "Edison and the Electric Light," *Manufacturer and Builder* 12. no. 2 (February 1880): 40.

19. *Nature* (12 February 1880): 342.

20. Charles Bazerman, *The Languages of Edison's Light* (Cambridge: MIT Press, 1999): 206.

21. "Edison's Light," New York *Herald*, 21 December 1879.

22. "Edison's Electric Light," *The New York Times*, 28 December 1879: 1.

23. Francis Upton to his father, 28 December 1879, Thomas Edison Papers Web site.

24. Henry A. Mott Jr., "Doubts about the electric light," *The New York Times*, 6 January 1880: 2.

25. "Edison's Electric Generator," *Telegraphic Journal and Electrical Review* (18 November 1879), Thomas Edison Papers Web site.

26. Friedel and Israel refer to the dynamo as Long-legged Mary Ann (70–72). In *Edison* (New York: McGraw-Hill, 1959), Matthew Josephson uses the epithet "Long-waisted Mary Ann" (244), but he does not document his source, and the phrase does not fit the configuration of the dynamo.

27. Attributed to the *Washington Journal*, 17 April 1880, at Thomas Edison Papers Web site.

28. *New York World*, 5 January 1880.

29. "Thomas A. Edison's Workshop," *The New York Times*, 16 January 1880: 1.

30. "Edison and the Electric Light": 342.

31. David Nye, *The Invented Self: An Anti-biography from Documents of Thomas Alva Edison* (Odense, Denmark: Odense University Press, 1983): 142.

32. "The Aldermen Visit Edison," *The New York Times*, 21 December 1880: 2.

33. Francis Jehl, "Only a Leak," *The Edison Monthly*, G.E. historical file, Schenectady Museum.

34. Edison's remarks, *Dinner in Honor of Thomas Alva Edison, Monday, the Eleventh of September 1922 at the Commodore Hotel* (New York: Edison Company, 1923).

35. Josephson: 261.

36. John Trowbridge, *Science* (29 February 1884): 261.

37. *Science* (10 May 1889): 367.

38. Robert Hammond, *The Electric Light in Our Homes* (London: Warner, 1884): 99.

39. Brian Bowers, *Lengthening the Day: A History of Lighting Technology* (Oxford: Oxford University Press, 1998): 127–28.

40. "Electric Lighting in New York," *Scientific American* (26 January 1884).

41. "Electricity at the Crystal Palace," *Nature* (9 March 1882): 446, 448.

42. "Electricity As a Factor in Happiness," *Appleton's Journal* 26 (1881): 467–69.

CHAPTER 4: NERVE JUICE

1. Edson to Edison, 26 October 1874, Thomas Edison Papers Web site.

2. Alphonso David Rockwell, *Rambling Recollections: An Autobiography* (New York: Paul B. Holbin, 1920): 188, 193. See also Barbara Sicherman, "The Paradox of Prudence: Mental Health in the Gilded Age," *Journal of American History* 62, no. 4 (March 1976): 901–5.

3. George Beard private journal, entry of April 1858, Yale University Manuscript and Archive Collection, Sterling Library: 78.

4. Ibid., 1859: 136; and November 1860: 154.

5. Ibid., March 1858: 73.

6. *The Twenty Years' Record of the Yale Class of 1862* (Bangor, ME: John H. Bacon, 1884): 24.

7. Mary Beard to George Beard, 15 July 1862: Yale University Manuscript and Archive Collection.

8. Spencer Beard to George Beard, 1 December 1856.

9. Ibid., Beard, private journal, 1862: 203.

10. Mary Beard to George Beard, 15 July 1862.

11. Beard, private journal, 1862: 188.

12. Ibid., 2 November 1866: 204.

13. George Miller Beard, "Biographical Record of Graduates," in *Decennial Meeting of the Class of 1862*, Yale. New Haven, 1973: 17.

14. Rockwell: 180, 192.

15. Beard private journal, February 1861: 157.

16. George Beard, *Our Home Physician* (New York: E. B. Treat, 1869): 600.

17. *Scientific American* (21 May 1870): 335.

18. "The Present Status of Medical Science," *Scientific American* (3 July 1869): 11.

19. Joseph Kett, *The Formation of the American Medical Profession: The Role of Institutions, 1780–1860* (New Haven: Yale University Press, 1968): 163.

20. Silas Weir Mitchell, *Fat and Blood: And How to Make Them* (Philadelphia: Lippincott, 1877): 43.

21. Suzanne Poirier, "The Weir Mitchell Rest Cure: Doctors and Patients," *Women's Studies* 10, no. 1 (1983): 15–40.

22. Beard, *Our Home Physician*: iii–iv, italics are in the original.

23. Helmholtz quoted in Charles Coulston Gillispie, *The Edge of Objectivity* (Princeton: Princeton University Press, 1960): 393.

24. Beard, *Our Home Physician*: 391.

25. William James to Henry James, 25 October 1869, *The Correspondence of William James*, vol. I, eds. Ignas K. Skrupskelis and Elizabeth M. Berkeley (Charlottesville: University Press of Virginia, 1992): 113.

26. See Iwan Rhys Morus, *Frankenstein's Children* (Princeton: Princeton University Press, 1998): 231–55.

27. Morton Prince, "The True Position of Electricity As a Therapeutic Agent in Medicine," *Boston Medical and Surgical Journal* 123, no. 14 (2 October 1890): 313.

28. My thanks to Ellen Kuhfeld, curator of instruments at the Bakken Library, for this information. See also Roberts Bartholow, *Medical Electricity* (Philadelphia: Lea, 1887): 87.

29. Eugene Taylor, "The electrified hand—psychotherapeutic implications," *Medical Instrumentation* 17, no. 4 (July–August 1983): 281–82.

30. William Begley Thorne to William Wilberforce Baldwin, 1 April 1905, Baldwin letters, Pierpont Morgan Library.

31. George Beard and Alphonso D. Rockwell, *A Practical Treatise on the Medical and Surgical Uses of Electricity* (New York: Wood, 1871): viii–xi.

32. Beard private journal, 1861: 183.

33. George Beard, "The Practice of Medicine in a Pecuniary Point of View," *The Medical Record* 3, no. 64 (15 October 1868): 372–75.

34. Beard and Rockwell: 301–3.

35. Edward Holmes Van Deusen, "Observations on a Form of Nervous Prostration (Neurasthenia) Culminating in Insanity," *American Journal of Insanity* 25 (1868–69): 445–61.

36. Beard and Rockwell: 299. Beard's first publication about neurasthenia, "Neurasthenia, or Nervous Exhaustion," appeared in the *Boston Medical and Surgical Journal* 80 (29 April 1869): 217–21.

37. George Beard, *Sexual Neurasthenia*, 5th ed., ed. A. D. Rockwell (New York: Arno, 1972 [New York: E. B. Treat, 1884, 1889]): 63, 64.

38. George Beard, "Who of Us Are Insane?," *Putnam's Monthly* 12, no. 11 (November 1868): 513–27.

39. Sigmund Freud, *The Freud Reader*, ed. Peter Gay (New York: Norton, 1995): 9–10.

40. Alphonso D. Rockwell, Preface to Beard and Rockwell: iii–iv.

41. Beard and Rockwell: 304, 170–71.

42. George Beard, "The Relation of the Soul and the Intellect to the Brain," *The Medical Record* (15 April 1869): 74–77.

43. Rockwell, *Rambling Recollections*: 183.

44. Rockwell to Charles L. Dana, quoted in Neil Baldwin, *Edison, Inventing the Century* (New York: Hyperion, 1995): 431.

CHAPTER 5: SPARKS

1. George Beard, "Experiments with the Alleged New Force" (New York: Clacher, 1876): 25.

2. Paul Israel, *Edison: A Life of Invention* (New York: Wiley, 1998): 94–95.

3. Beard, "Experiments with the Alleged New Force": 23.

4. Thomas Edison, *The Diary and Sundry Observations of Thomas A. Edison*, ed. Dagobert D. Runes (New York: Philosophical Library, 1948): 234.

5. George Parsons Lathrop, "Talks with Edison," *Harper's* (February 1890): 434–35.

6. P.C.B., "Phonomime, Autophone, and Kosmophone," *The New York Times*, 11 June 1878: 5.

7. Lathrop: 432.

8. Israel: 111.

9. John Cowan quoted in "A New Form of Electricity," *The Manufacturer and Builder* 8, no. 1 (January 1876): 14–15.

10. Beard, "Experiments with the Alleged New Force": 10.

11. Frank L. Dyer and Thomas C. Martin, *Edison, His Life and Inventions*, vol. 2 (New York: Harpers, 1910): 826–27.

12. Beard, "Experiments with the Alleged New Force": 12.

13. Edison letter dated 8 December 1875 to the editor of *Scientific American*, published 1 January 1876, in *The Papers of Thomas Alva Edison: The Making of an Inventor, February 1847–June 1873*, vol. II, eds. Reese V. Jenkins et al. (Baltimore: Johns Hopkins University Press, 1989): 680–81.

14. George Beard, "The Nature of the Newly Discovered Force," *Scientific American* (22 January 1876): 57.

15. Heaviside to Hertz, 14 February 1889, quoted in J. G. O'Hara and W. Pricha, *Hertz and the Maxwellians* (London: Peter Peregrinus Ltd, 1987): 58.

16. Beard, "Experiments with the Alleged New Force": 17.

17. Edison quoted in Matthew Josephson, *Edison* (New York: McGraw-Hill, 1959): 281.

18. "Edison in His Workshop," *Harper's Weekly* XXIII (2 August 1879); reprinted in *Popular Culture and Industrialism, 1865–1890*, ed. Henry Nash Smith (New York: New York University Press, 1967): 73–77.

19. For a study of the evolution of Edison's public image, see Wyn Wachhorst, *Thomas Alva Edison: An American Myth* (Cambridge: MIT Press, 1981).

20. "Edison's Electric Light," *Telegraphic Journal and Electrical Review* (1 October 1879), Thomas Edison Papers Web site.

21. William H. Bishop, "A Night with Edison," *Scribner's Monthly* 18 (November 1878): 98–99.

22. William James, "Is Life Worth Living?," in *The Will to Believe and other essays in popular philosophy* (New York: Dover, 1956 [1897, 1898]): 53.

23. Edison quoted in Israel: 260.

24. Horace Townsend, "Edison, His Work, and His Work-Shop," *Cosmopolitan* (April 1889): 598–607.

25. See Wachhorst: 33, 45, 92, 111, 116, 200.

PART II: CRAVINGS OF THE HEART

CHAPTER 6: THE INCONSTANT BATTERY

1. Homer Clark Bennett's comprehensive handbook, *An electrotherapeutic guide, or a thousand questions asked and answered* (Lima, OH: Literary Department of the National College of Electrotherapeutics, 1907), lists eighty-one ailments for which electrotherapy is appropriate: abscess, preventing or inciting abortion, acne, alcoholism, amenorrhea, anemia, to produce or remove anesthesia, angina, aphasia, apoplexy, arthritis, asphyxia, asthma, bust and penile development, muscular atrophy, baldness, blemishes, neuralgia, Bright's disease, bronchitis, carcinoma, catarrh, chilblains, conjunctivitis, constipation, consumption, cystitis, deafness, diabetes, diarrhea, dysmenorrhea, dyspepsia, eczema, seminal emissions, endometritis, enlarged prostate, epilepsy, erysipelas, various eye diseases, fever, fibroids, gastralgia, gastritis, goiter, gout, headaches, hemorrhoids, hiccough, hives, hypochondria, hysteria, impotence, incontinence, inflammation, insanity, jaundice, liver cirrhosis, liver spots, lumbago, lupus, melancholia, meningitis, menopause, obesity, paralysis, poliomyelitis, salpingitis, sarcoma, syphilis, tetanus, enlarged tonsils, tuberculosis, ulcer, uterine displacements, vaginal leucorrhoea, vaginismus, varicose veins, warts, wrinkles, writer's cramp.

2. Miss Caroline Smith, "A Prophetical Warning," Collection of the Bakken Library, n.d.

3. William James Morton, "Electricity in Medicine from a Modern Standpoint," *New York Medical Journal* 51 (20 April 1895): 488.

4. Oliver Wendell Holmes, *Medical Essays, 1842–1882* (Boston: Houghton Mifflin, 1892): 275, 276.

5. Hermann von Helmholtz, "On Thought in Medicine," in *Science and Culture: Popular and Philosophical Essays*, ed. David Cahan (Chicago: University of Chicago Press, 1995): 318. See also Malcolm Nicolson, "The Introduction of Percussion and Stethoscopy to Early-Nineteenth-Century Edinburgh" and Roy Porter, "The Rise of Physical Examination," in *Medicine and the Five Senses*, eds. W. F. Bynum and Roy Porter (Cambridge: Cambridge University Press, 1993): 134–53, 179–97.

6. Review of *Nature in Disease* by Jacob Bigelow, *North American Review* 80 (April 1855): 465.

7. Allan McLane Hamilton, *Recollections of an Alienist* (New York: Doran, 1916): 64.

8. "Bacteriology in Our Medical Schools," *Science,* 16 March 1888: 123–26.

9. *Scientific American* (14 June 1884).

10. See Margaret Fuller, *Woman in the Nineteenth Century* (New York: Norton, 1971 [1845]): 103–5, for a discussion of women's especially sensitive "magnetic element."

11. George Beard, *Sexual Neurasthenia*, 5th ed., ed. A. D. Rockwell (New York: Arno, 1972 [New York: E. B. Treat, 1889, 1884]): 61.

12. Beard made this point several times; for example, in his lecture "American Nervousness," given before the Philosophical Society of New York, 7 November 1880.

13. George Beard, *American Nervousness* (New York: Putnam, 1881): 105, 102–3.

14. Ibid.: 114.

15. Edward Reynolds, "On the Value of Electricity in Minor Gynecology," *Boston Medical and Surgical Journal* 126, no. 16 (21 April 1892): 383–85.

16. Alexander R. Becker, "Electricity as a Therapeutic Agent in Gynaecology," *Boston Medical and Surgical Journal* 104 (7 January 1886).

17. Beard, *Sexual Neurasthenia*: 26–27, 15, 225.

18. Gail Pat Parsons, "Equal Treatment for All: American Medical Remedies for Male Sexual Problems: 1850–1900," *Journal of the History of Medicine* (January 1977): 66.

19. Beard private journal, 2 November 1866, Yale University Manuscript and Archive Collection, Sterling Library: 205.

20. Charles L. Dana, "Dr. George M. Beard," *Archives of Neurology and Psychiatry* 10 (1923): 430.

21. *Minneapolis Tribune*, 21 April 1886: 8.

22. *St. Louis Post-Dispatch*, 14 February 1897: 14.

23. Roberts Bartholow, "What Is Meant by Nervous Prostration?," *Boston Medical and Surgical Journal* 90, no. 3 (17 January 1884): 53–56.

24. George Beard, "Medical Education and the Medical Professions in Europe and America," 1 December 1882, typescript, Beard papers, Yale University Manuscript and Archive Collection, Sterling Library.

25. Sigmund Freud, "Review of Averbeck's *Die Akute Neurasthenia*," quoted in M. B. Macmillan, "Beard's Concept of Neurasthenia and Freud's Concept of the Actual Neuroses," *Journal of the History of the Behavioral Sciences* 12, no. 4 (1976): 380.

26. Sigmund Freud to Wilhelm Fliess, 24 November 1887, in *The Complete Letters of Sigmund Freud to Wilhelm Fliess, 1887–1904*, trans. and ed. Jeffrey Moussaieff Masson (Cambridge: Belknap Press of Harvard University Press, 1985): 15.

27. Sigmund Freud, Draft A and Draft B of "The Etiology of the Neuroses," 1892, 1893, in *The Complete Letters of Sigmund Freud*: 42, 38.

28. Alphonso Rockwell, "The Late Dr. George M. Beard," (4 October 1883).

29. Hamlin Garland, "The Electric Lady," *Cosmopolitan* 29 (1900): 73–85.

30. James Maclaren Cobban, *The Master of His Fate* (Elstree, U.K.: Greenhill, 1987 [1890]): 84–85.

31. Ibid.: 58.

32. Ibid.: 86.

33. Ibid.: 217.

34. Ibid.: 219–20.

35. Ibid.: 138.

36. Ibid.: 141, 145.

37. Ibid.: 40, 157.

CHAPTER 7: HAUNTED BRAINS

1. Dods published nine books between 1850 and his death in 1872; two volumes appeared posthumously. Robert Collyer, *Mysteries of the vital element*, 2d ed. (London: Renshaw, 1971).

2. Martineau quoted in Valerie K. Pichanick, *Harriet Martineau, the Woman and Her Work* (Ann Arbor: University of Michigan Press, 1980): 130–31.

3. For more about Martineau and Victorian attitudes about mesmerism, see Alison Winter, *Mesmerized: Powers of Mind in Victorian Britain* (Chicago: University of Chicago Press, 1998).

4. Elizabeth Barrett to Robert Browning, 24–25 January 1846, in *Letters of Robert Browning and Elizabeth Barrett Browning, 1845–46*, ed. Elvan Kintner (Cambridge: Belknap Press of Harvard University Press, 1969): 416.

5. Ibid., Robert Browning to Elizabeth Barrett, 27 January 1846: 424.

6. Ibid., Harriet Martineau to Elizabeth Barrett, 8 February 1846: 461.

7. Fuller to Emerson, 28 January 1844, in *The Letters of Margaret Fuller*, vol. III, 1842–44, ed. Robert N. Hudspeth (Ithaca: Cornell University Press, 1984): 177–78.

8. Kristin Boudreau, *Sympathy in American Literature* (Gainesville: University Press of Florida, 2002): 8.

9. Edgar Allan Poe, "A Tale of the Ragged Mountains," in *The Science Fiction of Edgar Allan Poe*, ed. Harold Beaver (New York: Penguin, 1976): 99–109.

10. Ralph Waldo Emerson, *Journals of Ralph Waldo Emerson*, vol. VIII, eds. William Gilman and J. E. Parsons (Cambridge: Belknap Press of Harvard University Press, 1970): 341.

11. Poe, "Mesmeric Revelation," in *The Science Fiction of Edgar Allan Poe*: 124–34.

12. Poe to Chivers, 10 July 1844, in *Letters of Edgar Allan Poe*, vol I., ed. John Ward Ostrom (Cambridge: Harvard University Press, 1948): 260.

13. Poe, "The Facts in the Case of M. Valdemar," in *The Science Fiction of Edgar Allan Poe*: 393n, 394n.

14. Ibid.: 194–203.

15. Poe, "Some Words with a Mummy," in *The Science Fiction of Edgar Allan Poe*: 154–70.

16. See for example Chauncy Hare Townshend's *Facts in mesmerism, with reasons for a dispassionate inquiry into it*, 1841; *A Practical Magnetizer*, 1843; mesmerist, clairvoyant, and spiritualist Andrew Jackson Davis's *Lectures on clairmativeness or human magnetism*, 1845; Charles Morley's *Elements of Animal Magnetism*, 1847; George Barth's *Mesmerist's Manual of phenomena and practice: Directions for applying mesmerism to the cure of diseases*, 1852; and Dawson Bellhouse's *Ten minutes reading on medical galvanism*, 1855.

17. Angus McLaren, "Phrenology: Medium and Message," *The Journal of Modern History* 46, no. 1 (March 1974): 94, n. 40.

18. See "George Combe and the Remolding of Man's Constitution," in Roger Cooter, *The Cultural Meaning of Popular Science* (Cambridge: Cambridge University Press, 1984): 101–34.

19. George Eliot, *The Lifted Veil* (Edinburgh and London: Blackwood, n.d. [1878]): 275–341. The novella was written in 1859.

20. Simon Newcomb quoted in Albert E. Moyer, *A Scientist's Voice in American Culture: Simon Newcomb and the Rhetoric of Scientific Method* (Berkeley: University of California Press, 1992): 21–23.

21. Martineau quoted in McLaren: 94–95.

22. See Edward Hungerford, "Walt Whitman and His Chart of Bumps," *American Literature* 2, no. 4 (January 1931): 350–84; and Arthur Wroebel, "Whitman and the Phrenologists," *PMLA* 89 (January 1974): 17–23.

23. Poe to Frederick W. Thomas, 27 October 1841, *Letters of Edgar Allan Poe*: 185.

24. Twain quoted in Madeleine B. Stern, "Mark Twain Had His Head Examined," *American Literature* 41, no. 2 (May 1969): 208–9.

25. For considerations of the ambiguous meaning of sympathy, see Roy R. Male Jr., "Hawthorne and the Concept of Sympathy," *PMLA* 68 (1953): 138–49; Male, "Sympathy—a key word in American romanticism," *The Emerson Society Quarterly* 35 (1964): 19–23; Linda Simon, "Art and the Risks of Intimacy," *New England Quarterly* 72 (December 1999): 617–24; and Boudreau.

26. Charles L. Dana, "Dr. George M. Beard," *Archives of Neurology and Psychiatry* 10 (1923): 428.

27. George Beard, "The Psychology of Spiritism," *North American Review* 129, no. 272 (July 1879): 72.

28. See Edward M. Brown, "Neurology and Spiritualism in the 1870s," *Bulletin of the History of Medicine* 57 (1983): 563–77.

29. George Beard, "The Relation of the Soul and the Intellect to the Brain," *The Medical Record* (1869): 75.

30. Beard, "Experiments with the Alleged New Force" (New York: Clacher, 1876): 10–11.

31. Beard, "The Psychology of Spiritism": 67.

32. George Beard, "The Delusions of Clairvoyance," *Scribner's Monthly* 18, no. 3 (July 1879): 434–35.

33. Henry Olcott, *Inside the Occult: The True Story of Madame H. P. Blavatsky* (Philadelphia: Running Press, 1975): 8–9; reprint of *Old Diary Leaves: The True Story of the Theosophical Society* (New York: Putnam, 1895).

34. Thomas Edison, *The Diary and Sundry Observations of Thomas A. Edison*, ed. Dagobert D. Runes (New York: Philosophical Library, 1948): 233–35, 239–40.

35. Olcott: 466–67.

36. William James, "What Psychical Research Has Accomplished," in *The Will to Believe and other essays in popular psychology* (New York: Dover, 1956 [1897, 1898]): 300.

37. "Clairvoyance," *The New York Times*, 7 July 1881: 4.

38. David Nye, *The Invented Self: An Anti-biography from Documents of Thomas Alva Edison* (Odense, Denmark: Odense University Press, 1983): 151.

39. As Laura Otis notes in *Networking: Communicating with Bodies and Machines in the Nineteenth Century* (Ann Arbor: University of Michigan Press, 2001), the term *telepathy*, thinking at a distance, was analogous to *telegraphy*, writing at a distance: 182.

40. William James, "The Will," in *The Essential Writings*, ed. Bruce W. Wilshire (Albany: State University of New York Press, 1984): 42.

41. James, "Human Immortality," in *The Will to Believe*: 26–27.

42. F. W. H. Myers, "Glossary of Terms Used in Psychical Research," *The Proceedings of the Society for Psychical Research* 12 (1896–97): 166–74.

43. James, "What Psychical Research Has Accomplished," in *The Will to Believe*: 320.

CHAPTER 8: THE INSCRUTABLE SOMETHING

1. See *Oxford English Dictionary*, entry for *scientist*.

2. Ralph Waldo Emerson, entry of 13 July 1833, in *Emerson in His Journals*, ed. Joel Porte (Cambridge: Belknap Press of Harvard University Press, 1982): 110–11.

3. Ibid., entry of September 1848: 393.

4. William James, "Is Life Worth Living?," in *The Will to Believe and other essays in popular philosophy* (New York: Dover, 1956 [1897, 1898]): 40.

5. Ibid., "The Will to Believe": 21.

6. Ibid., "What Psychical Research Has Accomplished": 319–20.

7. Ibid.: 324, 325, 306.

8. Ibid., "The Will to Believe": 10.

9. Ibid., "Is Life Worth Living?": 51, 52.

10. William James, *Pragmatism* (Cambridge: Harvard University Press, 1978 [1906]): 55.

11. James, "The Will to Believe": 25–26.

12. Thomas C. Martin, "Nikola Tesla," *The Century* 477, no. 4 (February 1894): 584.

13. Ibid.: 584.

14. "Tesla's Lectures on Alternate Currents of High Potential and Frequency," *Nature* (11 February 1892): 345–46.

15. Robert Underwood Johnson, "In Tesla's Laboratory," *The Century* 49, no. 6 (April 1895): 933.

16. "Light without Heat," *The Manufacturer and Builder* 24, no. 5 (May 1892): 113.

17. Vere Withington, "An Electrical Study," *Overland Monthly* 20 (October 1892): 417–29. In *When Old Technologies Were New* (New York: Oxford University Press, 1988), Carolyn Marvin points out the prevalence in late-nineteenth-century fiction of "the lone woman forced by circumstances" to take a job made possible by the advent of electricity. "At this labor she captured the heart of a good man who wooed her from that unsheltered and risky occupation to become his wife" (27). Withington's story complicates this theme by presenting dire consequences of the man's vocation as scientist.

18. Gaby Wood considers the connection between Edison's talking doll and Villiers's *Tomorrow's Eve* in *Edison's Eve* (New York: Knopf, 2002): 111–63.

19. Auguste de Villiers de L'Isle-Adam, *Tomorrow's Eve*, trans. R. M. Adams (Urbana: University of Illinois Press, 1982 [1886]). In 1910, Edison donated $25 toward a memorial to Villiers.

20. Ibid., 61, 63.

21. Ibid., 64.

22. Ibid., 85.

23. Ralph Waldo Emerson, entry of 1848, in *Journals of Ralph Waldo Emerson*,

1847–1848, vol. X, ed. Merton M. Sealts Jr. (Cambridge: Belknap Press of Harvard University Press, 1973): 311.

24. For a full study of the assassination and its consequences, see Charles E. Rosenberg, *The Trial of the Assassin Guiteau* (Chicago: University of Chicago Press, 1968).

25. E. L. Godkin, "The Attempt on the President's Life," *The Atlantic Monthly* 48, no. 287 (September 1881): 398.

26. Charles L. Dana, "Expert Testimony," *North American Review* 138, no. 331 (June 1884): 613, 614.

27. George Beard, "Recent Works on the Brain and Nerves," *North American Review* 131 (September 1880): 282.

28. Gail Hamilton, "The Spent Bullet," *North American Review* 134 (May 1882): 532.

29. Ibid.: 527.

PART III: ELECTROSTRIKES

CHAPTER 9: LIVE WIRES

1. Alexandre Dumas, "The Slap of Charlotte Corday," in *Demons of the Night: Tales of the Fantastic, Madness and the Supernatural from Nineteenth-Century France*, ed. Joan C. Kessler (Chicago: University of Chicago Press, 1995): 145–71.

2. "Death by Electricity," *The New York Times*, 5 June 1888: 2.2.

3. Allan McLane Hamilton, *Recollections of an Alienist* (New York: Doran, 1916): 385.

4. "Death by Electricity," *The New York Times*, 17 January 1888: 8.1.

5. "Still Preferring Hanging," *The New York Times*, 9 March 1888: 5.4.

6. Peter H. Van Der Weyde, "The Comparative Dangers of Alternate vs. Direct Electric Currents," paper read before eighth convention of the National Electric Light Association, 29–31 August 1888.

7. "Offers Himself for Execution," *The New York Times*, 7 June 1889: 2.7.

8. "More Expert Testimony," *The New York Times*, 12 July 1889: 8.1.

9. "Power of Electricity," *The New York Times*, 16 July 1889: 8.1.

10. "Mr. Gerry on the Stand," *The New York Times*, 19 July 1889.

11. "Testimony of the Wizard," *The New York Times*, 24 July 1889: 21.1, 2.

12. "Execution by Electricity," *The New York Times*, 22 March 1890: 4.5.

13. Louis J. Palmer Jr., "In Re Kemmler," in *Encyclopedia of Capital Punishment in the United States* (Jefferson, NC: McFarland, 2001): 271.

14. "Kemmler's Last Night," *The New York Times*, 6 August 1890: 1.5.

15. "Death Is Drawing Nearer," *The New York Times*, 5 August 1890: 1.7.

16. "Far Worse Than Hanging," *The New York Times*, 7 August 1890: 1.5.

17. Newspapers quoted in "Kemmler's Terrible Death," *Western Electrician* (10 August 1890): 83–84.

18. Carlos Frederick MacDonald, *Report of Carlos F. MacDonald on the execution by electricity of William Kemmler alias John Hart* (Albany: Argus, 1890).

19. Hamilton: 383, 387–88.

20. Arthur Conan Doyle, "The Los Amigos Fiasco," *The Idler* (December 1892): 548–57.

21. *Nature* (4 April 1895): 539; *Nature* (11 April 1895): 562.

22. Mark Twain to Mary M. Fairbanks, 29 May 1891, in *Mark Twain to Mrs. Fairbanks*, ed. Dixon Wecter (San Marino, CA: Huntington Library, 1949): 267.

23. Arthur E. Kennelly, "Electricity in the Household," *Scribner's Monthly* 7 (1890): 102–15.

CHAPTER 10: MAGICAL KEYS

1. "Opened by the President," *The New York Times*, 2 May 1893: 1–3.

2. Henry Blake Fuller, "The Upward Movement in Chicago," quoted in Daniel Aaron's Introduction to Robert Herrick, *The Memoirs of an American Citizen* (Cambridge: Belknap Press of Harvard University Press, 1963): xi.

3. Henry Van Brunt, "The Columbian Exposition and American Civilization," *The Atlantic Monthly* 71, no. 427 (May 1893): 581.

4. Charles Moore, "Lessons of the Chicago World's Fair. An Interview with the Late Daniel H. Burnham," in *Architecture in America: A Battle of Styles*, eds. William A. Coles and Henry Hope Reed Jr. (New York: Appleton-Century-Crofts, [1961]): 146, 147.

5. Ibid.: 140.

6. Murat Halstead, "Electricity at the Fair," *Cosmopolitan* 15 (September 1893): 582.

7. Herrick: 147.

8. William James to Henry James, 22 September 1893, in *The Correspondence of William James*, vol. 2, eds. Ignas K. Skrupskelis and Elizabeth M. Berkeley (Charlottesville: University Press of Virginia, 1993): 280. William James did not attend the fair.

9. Charles M. Lungren, "Electricity at the World's Fair," *Popular Science Monthly* (October 1893): 721.

10. Halstead: 578.

11. Hubert Howe Bancroft, *Book of the Fair* (Chicago: Bancroft, 1893): 424.

12. Carolyn Marvin, *When Old Technologies Were New* (New York: Oxford University Press, 1988): 64.

13. Halstead: 578.

14. See Mary Gay Humphrey, "Talks with Decorators," *Art Amateur* 20, no. 5 (1889): 110–11; "Art and the Electric Light," *Art Amateur* 26, no. 2 (1892): 54–55; and Charles Bazerman, "The Language of Flowers: Domesticating Electric Light," in *The Languages of Edison's Light* (Cambridge: MIT Press, 1999).

15. Halstead: 577, 578.

16. Bancroft: 840, 836.

17. For a discussion of ethnology at the fair, see Robert W. Rydell, *All the World's a Fair* (Chicago: University of Chicago Press, 1984): 38–71.

18. Bancroft: 957.

19. Henry Adams, *The Education of Henry Adams* (New York: Vintage/Library of America, 1990 [1907]): 320, 317.

20. Norton to Fuller, 20 October 1893, in *Letters of Charles Eliot Norton*, vol. 2, eds. Sara Norton and M. A. De Wolfe Howe (Boston: Houghton Mifflin, 1913): 218.

21. "Editor's Study," *Harper's New Monthly Magazine* 87, no. 522 (November 1893): 960–61.

22. Dave Walter, ed., *Today Then: America's Best Minds Look 100 Years into the Future on the Occasion of the 1893 World's Columbian Exposition* (Helena, MT: American & World Geographic Publishing, 1992).

23. L. Frank Baum, *The Master Key: an electrical fairy tale* (Westport, CT: Hyperion, 1974 [1901]): 12.

24. Ibid.: 91, 93–95.

25. Ibid.: 86.

26. Ibid.: 236.

27. John Trowbridge, *The Electrical Boy, or, The Career of Greatman and Greatthings* (Boston: Roberts Brothers, 1891).

28. Trowbridge: 257.

29. Mark Twain, 7 October 1890, in *Mark Twain's Notebooks and Journals*, vol. III, eds. Robert Pack Browning, Michael Frank, and Lin Salamo (Berkeley: University of California Press, 1979): 548.

30. Mark Twain, "Mental Telegraphy," *Harper's Monthly* 22 (March 1891): 111–37; Twain, *Notebooks and Journals*: 347.

31. Mark Twain, *The American Claimant* (New York: Oxford University Press, 1996 [1892]): 45, 46, 81–82.

32. William Dean Howells, quoted in Linda Simon, "Looking Forward: A Profile of Edward Bellamy," *The World and I* (June 1999): 293.

33. Edward Bellamy, *Looking Backward, 2000–1887*, ed. Daniel H. Borus (Boston: Bedford Books of St. Martin's Press, 1995): 168. See also Borus's Introduction for historical and biographical context.

34. Edward Bellamy, "With the Eyes Shut," *Harper's New Monthly Magazine* 79, no. 473 (October 1889): 736–45.

35. Crookes quoted in "George Miller Beard," in *The Twenty Years' Record of the Yale Class of 1862* (Bangor, ME: John H. Bacon, 1884): 27.

36. Simon Newcomb, *His Wisdom, the Defender* (New York: Arno, 1975 [1900]): 314.

37. See Bruce Sinclair, "Technology on Its Toes: Late Victorian Ballets, Pageants, and Industrial Exhibitions," in *In Context: History and the History of Technology*, eds. Stephen H. Cutcliffe and Robert C. Post (Bethlehem: Lehigh University Press, 1989): 71–87.

CHAPTER 11: DARK LIGHT

1. *Science* 3, no. 56 (24 January 1896): 131.

2. Ibid.: 132; and *Science* 3, no. 57 (31 January 1896).

3. *Scientific American* (1 February 1896): 67; and "Roentgen or X-ray Photography," *Scientific American* (15 February 1896): 103.

4. *Philadelphia Press* to Edison, 7 February 1896; and Samuel McClure to Edison, requesting an update on his research, 11 February 1896, Edison Papers Web site.

5. Edison to Kennelly, 27 January 1896, quoted in Paul Israel, *Edison: A Life of Invention* (New York: Wiley, 1998): 309.

6. "Fluoroscope is a Success," *New York Herald*, 28 March 1896, Edison Papers Web site.

7. Thomas Commerford Martin, R. W. Wood, and Elihu Thomson, "Photographing the Unseen: A Symposium on the Roentgen Rays," *The Century* 52, no. 1 (May 1896): 125.

8. "Our Limited Vision and the New Photography," *The Lancet*, reprinted in *Littell's Living Age* 208 (28 March 1896): 821.

9. "Scientific Inquisitors," *The World*; reprinted in *Littell's Living Age* 208 (28 March 1896): 822–23.

10. David Starr Jordan, "The Sympsychograph: A Study in Impressionistic Physics," *Popular Science Monthly* 49 (September 1896): 597–602.

11. George Sarton, "The Discovery of X-Rays," *Isis* 26, no. 2 (March 1937): 356.

12. Martin, Wood, and Thomson: 128.

13. James D. Newton, *Uncommon Friends* (New York: Harcourt, Brace, Jovanovich, 1987): 21.

14. Charles R. R. Hayter, "The Clinic As Laboratory: The Case of Radiation Therapy, 1896–1920," *Bulletin of the History of Medicine* 72, no. 4 (1998): 675–76.

15. Homer Clark Bennett, *An electrotherapeutic guide, or a thousand questions asked and answered* (Lima, OH: Literary Department of the National College of Electrotherapeutics, 1907): 235.

16. Paul Dubois, *Self-Control and How to Secure It*, trans. Harry H. Boyd (New York: Funk & Wagnalls, 1909): 4.

17. Paul Dubois, *The Psychic Treatment of Nervous Disorders*, trans. Smith Ely Jelliffe and William A. White (New York: Funk & Wagnalls, 1905): 210.

18. William James, *The Varieties of Religious Experience* (Cambridge: Harvard University Press, 1985): 84, 86, 102.

19. William James, "Address on the Medical Registration Bill," 2 March 1898, in *The Works of William James: Essays, Comments, and Reviews* (Cambridge: Harvard University Press, 1987): 56–62.

20. James, *Varieties*: 381, 192n.

21. Sigmund Freud and Josef Breuer, *Studies on Hysteria. The Standard Edition of the Complete Psychological Works of Sigmund Freud*, vol. II, trans. James Strachey (London: Hogarth, 1955): 7; italics are in the original. The patient quoted here is Anna O., who had been treated by Breuer.

22. Freud to Wilhelm Fliess, 1 January 1896, *The Complete Letters of Sigmund Freud to Wilhelm Fliess, 1887–1904*, trans. and ed. Jeffrey Moussaieff Masson (Cambridge: Belknap Press of Harvard University Press, 1985): 162–69.

23. Freud and Breuer: 17; italics are in the original.

24. Ibid.: 294–95.

25. Ibid.: 136.

26. William James, "What Psychical Research Has Accomplished," in *The Will to Believe and other essays in popular philosophy* (New York: Dover, 1956 [1897, 1898]): 327.

27. James Jackson Putnam, "Recent Experiences in the Study and Treatment of Hysteria at the Massachusetts General Hospital; with Remarks on Freud's Method of Treatment by 'Psycho-Analysis,'" *Journal of Abnormal Psychology* (April 1906): 36, 40–41.

28. See Max Nordau, *Degeneration* (New York: Howard Fertig, 1968 [1892, trans. 1895]). Nordau, a physician, saw mysticism, pessimism, impulsiveness, and egoism as unfortunate symptoms of his age. His strident characterization of his contemporaries' obsessive self-analysis generated equally strident criticism. Nordau replied in "A Reply to My Critics," *The Century* 50, no. 4 (August 1895): 546–51.

29. August Strindberg, *By the Open Sea*, trans. Ellie Schleussner (New York: Haskell House, 1973 [1890]): 74–75.

30. Ibid.: 75.

31. Ibid.: 121.

32. Ibid.: 183.

33. Ibid.: 185.

34. Ibid.: 316.

35. Henry Adams, *The Education of Henry Adams* (New York: Vintage/Library of America: 1990 [1907]): 354, 355.

36. Rudyard Kipling, "Wireless," in *Traffics and Discoveries* (New York: Scribner's, 1904 [1902]): 239–68.

37. Robert Thurston, "The Animal As a Machine," *North American Review* 163, no. 480 (November 1896): 607–19.

38. "An Artificial Man," *American Monthly Review of Reviews* (July 1906): 120.

Bibliography

Adams, Henry. *The Education of Henry Adams*. New York: Vintage/Library of America, 1990 (1907).

Adams, Judith. "The Promotion of New Technology Through Fun and Spectacle: Electricity at the World's Columbian Exposition." *Journal of American Culture* 18, no. 2 (1995): 45–55.

Altschule, Mark D. "Ideas of Eighteenth-Century British Medical Writers about Anxiety." *New England Journal of Medicine* 248 (9 April 1953): 646–53.

Armstrong, Tim. *Modernism, Technology and the Body*. Cambridge: Cambridge University Press, 1998.

Arnett, L. D. "The soul—a study of past and present beliefs." *American Journal of Psychology* 15 (1904): 121–200, 347–82.

"Art and the electric light." *Art Amateur* 26, no. 2 (1892): 54–55.

Baldick, Chris. *In Frankenstein's Shadow: Myth, Monstrosity, and Nineteenth-Century Writing*. Oxford: Clarendon, 1987.

Baldwin, Neil. *Edison, Inventing the Century*. New York: Hyperion, 1995.

Barnard, Charles, "Electricity, something it is doing." *The Century* 37, no. 5 (March 1889): 736–41.

Barrett, J. P. *Electricity at the Columbian Exposition*. Chicago: Donnelly, 1894.

Barrett, William, Edmund Gurney, and F. W. H. Myers, "First Report of the Committee on Mesmerism." *The Proceedings of the Society for Psychical Research* 1 (1882–83): 220.

Bartholow, Roberts. *Medical Electricity*. Philadelphia: Lea, 1887.

———. "What Is Meant by Nervous Prostration?" *Boston Medical and Surgical Journal* 90, no. 3 (17 January 1884): 53–56.

Bassuk, Ellen. "The Rest Cure: Repetition or Resolution of Victorian Women's Conflicts." In *The Female Body in Western Culture*. Edited by Susan Suleiman. Cambridge: Harvard University Press, 1986: 139–51.

Baum, L. Frank. *The Master Key: an electrical fairy tale*. Westport, CT: Hyperion, 1974 (1901).

Bayerz, Kurt. "Biology and Beauty: Science and Aesthetics in *Fin-de-Siècle* Germany." In *Fin-de-Siècle and Its Legacy*. Edited by Mikulas Teich and Roy Porter. Cambridge: Cambridge University Press, 1990: 278–95.

Bazerman, Charles. *The Languages of Edison's Light*. Cambridge: MIT Press, 1999.

Beard, George. *American Nervousness*. New York: Putnam, 1881.

———. "The Delusions of Clairvoyance." *Scribner's Monthly* 18, no. 3 (July 1879): 433–40.

———. "The Elements of Electrotherapeutics." *Archives of Electrology and Neurology* 1, no. 1 (1874): 158–64; 1, no. 2 (1874): 17–23, 184–94.

———. "English and American Physique." *North American Review* 129, no. 277 (December 1879): 588–603.

———. "Experiments with Living Human Beings." *Popular Science Monthly* 14: 611–21, 751–57.

———. *Herbert Spencer on American Nervousness: A Scientific Coincidence*. New York: Putnam, 1882.

———. "The Moral Responsibility of the Insane." *North American Review* 134, no. 302 (January 1882): 11–17.

———. "Neurasthenia, or Nervous Exhaustion." *Boston Medical and Surgical Journal* 80 (29 April 1869): 217–21.

———. "The Newly-Discovered Force." *Archives of Electrology and Neurology* 1, no. 2 (1874): 209–16.

———. *Our Home Physician*. New York: E. B. Treat, 1869.

———. "Physical Future of the American People." *The Atlantic Monthly* 43 (June 1879): 718–28.

———. "The Psychology of Spiritism." *North American Review* 129, no. 272 (July 1879): 65–81.

———. "Recent Works on the Brain and the Nerves." *North American Review* 131 (September 1880): 278–84.

———. "The Scientific Study of Human Testimony." *Popular Science Monthly* 13 (May 1878): 53–64; (June): 173–83; (July): 328–38.

———. *Sexual Neurasthenia*, 5th ed. Edited by A. D. Rockwell. New York: Arno, 1972 (New York: E. B. Treat, 1884, 1889).

———. "Who of Us Are Insane?" *Putnam's Monthly* 12, no. 11 (November 1868): 513–27.

———. "A Year of Experiment in Electrotherapeutics." Paper read before the New York Academy of Medicine, 16 May 1872. Louisville: J. P. Morton, 1872.

Beard, George, and Alphonso D. Rockwell. *A Practical Treatise on the Medical and Surgical Uses of Electricity*. New York: Wood, 1871.

Becker, Robert O., and Gary Selden. *The Body Electric, Electromagnetism and the Foundation of Life*. New York: Morrow, 1985.

Bedini, Silvio. "Automata in History." *Technology and Culture* 5, no. 1 (Winter 1964): 24–42.

Beer, Gillian. *Darwin's Plots: Evolutionary Narrative in Darwin, George Eliot, and Nineteenth Century Fiction*. New York: Routledge, 1983.

Beichman, Arnold. "The First Electrocution." *Commentary* 35 (May 1963): 410–19.

Bellamy, Edward. *Looking Backward, 2000–1887*. Edited by Daniel H. Borus. Boston: Bedford Books of St. Martin's Press, 1995.

———. "With the Eyes Shut." *Harper's New Monthly Magazine* 79, no. 473 (October 1889): 736–45.

Bender, Bert A. "Let There Be (Electric) Light! The Image of Electricity in American Writing." *Arizona Quarterly* 34, no. 1 (1978): 55–70.

Benjamin, Park. *The Age of Electricity from amber-soul to telephone*. New York: Scribner's, 1888.

———. *The Intellectual Rise in Electricity: a history*. New York: Appleton, 1895.

Benjamin Franklin's Experiments. Edited by I. Bernard Cohen. Cambridge: Harvard University Press, 1941.

Bennett, Homer Clark. *An electrotherapeutic guide, or a thousand questions asked and answered*. Lima, OH: Literary Department of the National College of Electrotherapeutics, 1907.

Benton, E. "Vitalism in Nineteenth-Century Scientific Thought." *Studies in the History and Philosophy of Science* 5 (1974): 17–48.

Berliner, Emile. "The Gramophone: Etching the Human Voice." *Journal of the Franklin Institute* 125 (22 June 1888): 425–47.

Bigelow, Horatio R., ed. *An International System of Electrotherapeutics*. Philadelphia: Davis, 1894.

Bishop, Dr. Francis B. "On the Static Cage and Its Uses." *Transactions of the American Electrotherapy Association* (1896): 108–12.

Bishop, William H. "A Night with Edison." *Scribner's Monthly* 18 (November 1878): 98–99.

Blondheim, Menachem. *News Over the Wires*. Cambridge: Harvard University Press, 1994.

———. "When Bad Things Happen to Good Technologies: Three Phases in the Diffusion and Perception of American Telegraphy." In *Technology, Pessimism, and Postmodernism*. Edited by Yaron Ezrahi, Everett Mendelsohn, and Howard P. Segal. Amherst: University of Massachusetts Press, 1994: 77–92.

Bowditch, Henry P. "What Is Nerve-Force?" Address to American Association of the Advancement of Science, Buffalo, August 1886. Salem, MA: Salem Press, 1886.

Bowers, Brian. *Lengthening the Day: A History of Lighting Technology*. Oxford: Oxford University Press, 1998.

Brackett, C. F. "Electricity in the Service of Man." *Scribner's Monthly* 5, no. 6 (June 1889): 643–59.

Brandon, Craig. *The Electric Chair: An Unnatural American History*. Jefferson, NC: McFarland, 1999.

Brantlinger, Patrick. "Mass Media and Culture in *Fin-de-Siècle* Europe." In *Fin-de-Siècle and Its Legacy*. Edited by Mikulas Teich and Roy Porter. Cambridge: Cambridge University Press, 1990: 98–113.

Brazier, Mary. *A History of Neurophysiology in the Nineteenth Century*. New York: Raven, 1988.

Britten, Emma Hardinge. *The Electric Physician: or Self Cure through Electricity*. Boston: W. Britten, 1875.

Brown, Edward M. "Neurology and Spiritualism in the 1870s." *Bulletin of the History of Medicine* 57 (1983): 563–77.

Brown, Harold P. "The New Instrument of Execution." *North American Review* (November 1889): 587–93.

Brown, Theodore M. "From Mechanism to Vitalism in Eighteenth-Century Physiology." *Journal of the History of Biology* 7 (1974): 179–216.

Brush, Charles. *The Brush System of Electric Lighting*. Cleveland: Wiseman and Harvey, 1879.

———. "Some Reminiscences of Early Electric Lighting." 18 April 1928. Hammond Historical File, Schenectedy Museum.

Bunker, Henry Alden, Jr. "From Beard to Freud: A Brief History of the Concept of Neurasthenia."*Medical Review of Reviews* 36 (1930): 108–14.

Burg, David F. *Chicago's White City of 1893*. Lexington: University Press of Kentucky, 1976.

Burr, Anne Robeson. *Weir Mitchell: His Life and Letters*. New York: Duffield, 1929.

Bynner, Edwin L. "Diary of a Nervous Invalid." *The Atlantic Monthly* 71 (1893): 33–46.

Bynum, W. F., and Roy Porter, eds. *Companion Encyclopedia of the History of Medicine*, 2 vols. New York: Routledge, 1993.

———. *Medical Fringe and Medical Orthodoxy, 1750–1850*. London: Croom Helm, 1987.

Bynum, W. F., Roy Porter, and Michael Shepherd, eds. *The Anatomy of Madness*, vol. 1. London: Tavistock, 1985.

Cameron, Sharon. *The Corporeal Self: Allegories of the Body in Melville and Hawthorne*. Baltimore: Johns Hopkins University Press, 1981.

Cantor, Geoffrey N. *Michael Faraday: Sandemanian and Scientist*. New York: St. Martin's, 1991.

Cantor, Geoffrey N., and M. J. S. Hodge, eds. *Conceptions of Ether: Studies in the History of Etheric Theories, 1740–1900*. Cambridge: Cambridge University Press, 1981.

Caplan, Eric. *Mind Games: American Culture and the Birth of Psychotherapy*. Berkeley: University of California Press, 1998.

Carlson, Eric T. "The Nerve Weakness of the 19th Century." *International Journal of Psychiatry* 9 (1970): 50–54.

Carpenter, William B. "On the Mutual Relations of the Vital and Physical Forces." *Philosophical Transactions of the Royal Society of London* 140 (1950): 727–57.

Cassedy, James H. *Medicine in America: a short history*. Baltimore: Johns Hopkins University Press, 1991.

Channell, David F. *The Vital Machine: A Study of Technology and Organic Life*. Oxford: Oxford University Press, 1991.

Channing, W. F. *Notes on the Medical Applications of Electricity*. Boston: D. Davis, 1849.

Clarke, Edwin, and L. S. Jacyna. *Nineteenth-Century Origins of Neuroscientific Concepts*. Berkeley: University of California Press, 1987.

Cleaves, Margaret A. *An Autobiography of a Neurasthene. As Told by One of Them and Recorded by Margaret A. Cleaves*. Boston: Gorham, 1910.

———. *Light Energy, its physics, physiological action and therapeutic applications*. New York: Rebman, 1904.

———. *The record of four years (1895–99) in an exclusively electro-therapeutic clinic*. New York: [n.p.], 1899.

Clower, William T. "The Transition from Animal Spirits to Animal Electricity: a neuroscience paradigm shift." *Journal of the History of the Neurosciences* 7, no. 3 (December 1998): 201–18.

Cobban, James Maclaren. *Master of His Fate*. Elstree, U.K.: Greenhill, 1987 [1890].

Cohen, I. Bernard. *Benjamin Franklin's Science*. Cambridge: Harvard University Press, 1990.

———. *Revolution in Science*. Cambridge: Belknap Press of Harvard University Press, 1985.

Cosslett, Tess. *The 'Scientific Movement' and Victorian Literature*. New York: St. Martin's, 1982.

Cowan, Ruth. *A Social History of American Technology*. Oxford: Oxford University Press, 1997.

Crabtree, Adam. *From Mesmer to Freud*. New Haven: Yale University Press, 1993.

Craig, Alexander. *Ionia: Land of Wise Men and Fair Women*. New York: Arno Press and The New York Times, 1971 (1898).

Cunningham, Andrew, and Nicholas Jardine, eds. *Romanticism and the Sciences*. Cambridge: Cambridge University Press, 1990.

The Curtain Lifted by the Electro Absorbent Appliance Company. Rockford, IL: A. F. Judd, 1884.

Curti, Merle. "America and the World Fairs, 1851–1893." *American Historical Review* 55, no. 4 (July 1950): 833–56.

Cutcliffe, Stephen H., and Robert C. Post. *In Context: History and the History of Technology*. Bethlehem: Lehigh University Press, 1989.

Dana, Charles L. "Dr. George M. Beard." *Archives of Neurology and Psychiatry* 10 (1923): 428–35.

———. "Expert Testimony," *North American Review* 138, no. 331 (June 1884): 613–17.

Darnton, Robert. *Mesmerism and the End of the Enlightenment in France*. Cambridge: Harvard University Press, 1968.

Daston, Lorraine J. "British Responses to Psycho-Physiology, 1860–1900." *Isis* 69, no. 2 (June 1978): 192–208.

Davenport, F. H. "Some Gynecological Cases Treated by the Faradic Current." *Boston Medical and Surgical Journal* CXIX, no. 17: 397–99.

Dawson, Paul. "'A Sort of Natural Magic': Shelley and Animal Magnetism." *Keats-Shelley Review* 1(1986): 15–34.

Dibner, Bern. *Early Electrical Machines*. Norwalk, CT.: Burndy Library, 1957.

Dickson, Antonia, and W. K. L. Dickson. "Edison's Invention of the Kineto-Phonograph." *The Century* 48, no. 2 (June 1894): 207–15.

Dinner in Honor of Thomas Alva Edison, Monday, the Eleventh of September 1922 at the Commodore Hotel. New York: Edison Company, 1923.

Doane, Mary Ann. "Technophilia: Technology, Representation, and the Feminine." In *Body/Politics: Women and the Discourses of Science*. Edited by Mary Jacobus, E. F. Keller, and Sally Shuttleworth. New York: Routledge, 1980.

Dods, John Bovee. *The Philosophy of Electrical Physiology*. New York: Fowler & Wells, 1850.

———. *The Philosophy of Mesmerism*. Boston: William Hall, 1843.

———. *Thirty Short Sermons, Both Doctrinal and Practical*. Boston: Whittemore, 1842.

Doyle, Sir Arthur Conan. "The Los Amigos Fiasco." *The Idler* (December 1892): 548–57.

Dresser, Horatio W. *Health and the Inner Life*. New York: Putnam, 1906.

Drinka, G. G. *The Birth of Neurosis: Myth, Malady, and the Victorians*. New York: Simon & Schuster, 1984.

Drinkwater, H. *Fifty Years of Medical Progress, 1873–1922*. New York: Macmillan, 1924.

Dubois, Paul. *The Psychic Treatment of Nervous Disorders*. Translated by Smith Ely Jelliffe and William A. White. New York: Funk & Wagnalls, 1905.

———. *Self-Control and How to Secure It*. Translated by Harry H. Boyd. New York: Funk & Wagnalls, 1909.

DuBois-Reymond, Emil. "The Seven World-Problems." *Popular Science Monthly* 20 (1881–82): 433–47.

Dyer, Frank L., and Thomas C. Martin. *Edison: His Life and Inventions*, 2 vols. New York: Harper & Bros., 1910.

Earnest, Ernest P. *S. Weir Mitchell, Novelist and Physician*. Philadelphia: University of Pennsylvania Press, 1950.

Edison, Thomas. "The Dangers of Electric Lighting." *North American Review* 149 (November 1889): 625–34.

———. *The Diary and Sundry Observations of Thomas A. Edison*. Edited by Dagobert D. Runes. New York: Philosophical Library, 1948.

———. *The Papers of Thomas Alva Edison: The Making of an Inventor, February 1847–June 1873*, 4 vols. Edited by Reese V. Jenkins et al. Baltimore: Johns Hopkins University Press, 1989.

——. "The Perfected Phonograph." *North American Review* 146 (June 1888): 641–50.

——. "The Phonograph and Its Future." *North American Review* 126 (May–June 1878): 527–36.

——. "The Success of the Electric Light." *North American Review* 131 (October 1880): 295–300.

"The Electrical Execution." *Electrical Review* (24 November 1888).

"Electricity As a Factor in Happiness." *Appleton's Journal* 26 (1881): 467–69.

"Electricity in the Household." *Scientific American* (19 March 1904): 232.

Eliot, George. *The Lifted Veil*. Edinburgh and London: Blackwood, n.d. [1878].

Ellenberger, Henri. *The Discovery of the Unconscious*. New York: Basic Books, 1970.

Evans, Richard. "Shocking Improvements: Electricity in the American Household at the Turn of the Century." *Nineteenth Century* 20, no. 1 (Spring 2000): 29–34.

"The Execution by Electricity." *Scientific American* (27 September 1890): 200.

"The Execution of Criminals by Electricity." *The Manufacturer and Builder* 21, no. 4 (April 1889): 79–80.

Ezrahi, Yaron, Everett Mendelsohn, and Howard P. Segal, eds. *Technology, Pessimism, and Postmodernism*. Amherst: University of Massachusetts Press, 1994.

Falk, Doris V. "Poe and the Power of Animal Magnetism." *PMLA* 84 (1969): 536–46.

Fara, Patricia. "An Attractive Therapy: Animal Magnetism in Eighteenth-Century England." *History of Science* 33, pt. 2, 100 (June 1995): 127–77.

Fellman, A. C., and M. Fellman. *Making Sense of Self. Medical Advice Literature in Late Nineteenth-Century America*. Philadelphia: University of Pennsylvania Press, 1981.

Finger, Stanley. *Origins of Neuroscience: A History of Explorations into Brain Function*. New York: Oxford, 1994.

Finger, Stanley, and Mark B. Law. "Karl August Weinhold and His 'Science' in the Era of Mary Shelley's Frankenstein: Experiments on Electricity and the Restoration of Life." *Journal of the History of Medicine* 53 (April 1998): 161–80.

Finn, Bernard S. "The Incandescent Electric Light." In *Bridge to the Future: A Centennial Celebration of the Brooklyn Bridge*. Edited by Margaret Latimer, Brooke Hindle, and Melvin Kranzberg. New York: New York Academy of Sciences, 1984: 247–63.

"The First Electrical Execution." *Scientific American* (16 August 1890): 96.

Fisch, Menachem. *William Whewell, Philosopher of Science*. Oxford: Clarendon, 1991.

Foster, George G. *New York by Gas-Light*. Berkeley: University of California Press, 1990 (1850).

Foucault, Michel. *The Birth of the Clinic*. Translated by A. M. Sheridan Smith. New York: Vintage, 1975 (1973).

——. *Madness and Civilization*. Translated by Richard Howard. New York: Mentor, 1966.

Francesco, Grete de. *The Power of the Charlatan*. New Haven: Yale University Press, 1939.

French, Roger K. "Ether-Physiology." In Geoffrey N. Cantor and M. J. S. Hodge, eds. *Conceptions of Ether*. (Cambridge: Cambridge University Press, 1981): 111–32.

Friedel, Robert, and Paul Israel. *Edison's Electric Light: Biography of an Invention*. New Brunswick: Rutgers University Press, 1986.

Frohock, Fred M. *Healing Powers: Alternative Medicine, Healing Communities, and the State*. Chicago: University of Chicago Press, 1992.

Freud, Sigmund, and Josef Breuer. *Studies on Hysteria. The Standard Edition of the Complete Psychological Works of Sigmund Freud*, vol. II. Translated by James Strachey. London: Hogarth, 1955.

Fuller, Robert C. *Alternative Medicine and American Religious Life*. New York: Oxford, 1989.

———. *Mesmerism and the American Cure of Souls*. Philadelphia: University of Pennsylvania Press, 1982.

Gabler, Edwin. *The American Telegrapher: A Social History, 1860–1900*. New Brunswick: Rutgers University Press, 1988.

Garland, Hamlin. "The Electric Lady," *Cosmopolitan* 29 (1900): 73–85.

Garrison, Fielding H. *An Introduction to the History of Medicine*. Philadelphia: Saunders, 1917.

Gay, Peter. *The Bourgeois Experience, Victoria to Freud*, vol. 2: *The Tender Passion*. New York: Oxford University Press, 1986.

Geison, Gerald. "The Protoplasmic Theory of Life and the Vitalist-Mechanist Debate." *Isis* 60 (1969): 273–92.

Gibson, Jane Mork. "The International Electrical Exhibition of 1884." *IEEE, Transactions on Education* E-23 (1980): 169–76.

Gillispie, Charles Coulston. *The Edge of Objectivity*. Princeton: Princeton University Press, 1960.

Glendenning, Victoria. *Electricity*. Boston: Little, Brown, 1995.

Goldstein, Jan. *Console and Classify: The French Psychiatric Profession in the 19th Century*. Cambridge: Cambridge University Press, 1987.

Good, Byron, and Mary-Jo Good. "The Meaning of Symptoms." In *The Relevance of Social Science to Medicine*. Edited by L. Eisenberg and A. Kleinman. Dordrecht, Holland: D. Reidel, 1981: 166–96.

Good, Mary-Jo, Del Vecchio, Paul E. Brodwin, Byron J. Good, and Arthur Kleinman, eds. *Pain As Human Experience: An Anthropological Perspective*. Berkeley: University of California Press, 1992.

Gooding, David, and Frank A. J. L. James, eds. *Faraday Rediscovered: Essays on the Life and Work of Michael Faraday*. London: Macmillan, 1985.

Gooding, David, Trevor Pinch, and Simon Schaffer, eds. *The Uses of Experiment*. Cambridge: Cambridge University Press, 1989.

Gorowitz, Bernard, et al., eds. *The General Electric Story*. Schenectady Museum Association, 1999.

Gosling, Francis. *Before Freud: Neurasthenia and the American Medical Community, 1870–1910*. Urbana: University of Illinois Press, 1987.

———. "Neurasthenia in Pennsylvania: A Perspective on the Origin of American Psychotherapy, 1870–1910." *Journal of the History of Medicine* 40 (1985): 188–206.

Gosling, Francis, and J. M. Ray. "The Right to Be Sick: American Physicians and Nervous Patients." *Journal of Social History* 20 (1986): 251–67.

Gower, Barry. "Speculation in Physics: The History and Practice of Naturphilosophie." *Studies in the History and Philosophy of Science* 3 (1973): 301–56.

"The Gramophone." *The Manufacturer and Builder* 20, no. 2 (1 February 1888): 39.

"The Gramophone." *The Manufacturer and Builder* 20, no. 8 (August 1888): 175–78.

Grant, A. Cameron. "Combe on Phrenology and Free Will: A Note on XIXth-Century Secularism." *Journal of the History of Ideas* 26, no. 1 (January–March 1965): 141–47.

Gray, B. M. "Pseudoscience and George Eliot's 'The Lifted Veil.'" *Nineteenth-Century Fiction* 36, no. 4 (March 1982): 407–23.

Greenberg, Mark L., and Lance Schachterle, eds. *Literature and Technology*. Bethlehem: Lehigh University Press, 1992.

Greenway, John L. "'Nervous Disease' and Electric Medicine." In *Pseudoscience and Society in Nineteenth-Century America*. Edited by Arthur Wrobel. Louisville: University of Kentucky Press, 1987: 46–73.

Grier, Matthew J. *The treatment of some forms of sexual debility by electricity*. Philadelphia: Medical Press, 1891.

Griggs, Earl Leslie, ed. *Collected Letters of Samuel Taylor Coleridge*, vols. I and II. Oxford: Clarendon, 1956.

Grob, Gerald N. "Psychiatry's Holy Grail: The Search for the Mechanisms of Mental Diseases." *Bulletin of the History of Medicine* 72, no. 2 (1998): 189–219.

Hale, Nathan G., Jr. *Freud and the Americans*. Oxford: Oxford University Press, 1971.

Haley, Bruce. *The Healthy Body and Victorian Culture*. Cambridge: Harvard University Press, 1978.

Hall, Thomas S. *Ideas of Life and Matter*, vol. 2. Chicago: University of Chicago Press, 1969.

Haller, John S., Jr. *American Medicine in Transition, 1840–1910*. Urbana: University of Illinois Press, 1981.

Haller, John S., Jr., and Robin Haller. *The Physician and Sexuality in Victorian America*. Urbana: University of Illinois Press, 1974.

Halstead, Murat. "Electricity at the Fair." *Cosmopolitan* 15 (September 1893): 577–83.

Hamilton, Allan McLane. *Recollections of an Alienist*. New York: Doran, 1916.

Hammond, Robert. *The Electric Light in Our Homes*. London: Warner, 1884.

Harman, P. M. *Energy, Force, and Matter: The Conceptual Development of Nineteenth-Century Physics*. Cambridge: Cambridge University Press, 1982.

Harrington, Anne. "A Feeling for the 'Whole': The Holistic Reaction in Neurology from the *Fin-de-Siècle* to the Interwar Years." In *Fin-de-Siècle and Its Legacy*.

Edited by Mikulas Teich and Roy Porter. Cambridge: Cambridge University Press, 1990: 254–77.

———. *Medicine, Mind, and the Double Drain: A Study in Nineteenth-Century Thought*. Princeton: Princeton University Press, 1989.

Hawthorne, Nathaniel. *The House of the Seven Gables*. New York: Penguin, 1986 (1851).

———. *Young Goodman Brown and Other Short Stories*. New York: Dover, 1992.

Hayter, Charles R. R. "The Clinic As Laboratory: The Case of Radiation Therapy, 1896–1920." *Bulletin of the History of Medicine* 72, no. 4 (1998): 663–88.

Heilbron, J. L. "The Contributions of Bologna to Galvanism." *Historical Studies in the Physical and Biological Sciences* 22, no. 1 (1991): 57–85.

———. *Electricity in the 17th and 18th Centuries: A Study of Early Modern Physics*. Berkeley: University of California Press, 1979.

———. *Elements of Early Modern Physics*. Berkeley: University of California Press, 1982.

Heimann, P. M. "The *Unseen Universe*: physics and philosophy of nature in Victorian Britain." *British Journal for the History of Science* 6 (1972): 73–79.

Hein, Hilde. "The Endurance of the Mechanism-Vitalism Controversy." *Journal of the History of Biology* 5 (1972): 159–88.

Helmholtz, Hermann von. "On the Modern Development of Faraday's Conception of Electricity." *Science* 2, no. 43 (23 April 1881): 182–85.

———. "On Thought in Medicine." In *Science and Culture: Popular and Philosophical Essays*. Edited by David Cahan. Chicago: University of Chicago Press, 1995: 309–27.

Hering, D. W. "A Year of the X-Rays." *Popular Science Monthly* 50 (1896): 654–62.

Hoffmann, Charles. "The Depression of the Nineties." *Journal of Economic History* 16, no. 2 (June 1956): 137–64.

Holcomb, Harry S. "Electrotherapy: A Wood Engraving." *Journal of the History of Medicine* 22, no. 2 (April 1967): 180–81.

Holland, Henry. *Recollections of Past Life*. London: Longmans, Green, 1872.

Holmes, Oliver Wendell. *Elsie Venner*. Boston: Houghton Mifflin, 1891 (1861).

———. *Medical Essays, 1842–1882*. Boston: Houghton Mifflin, 1892.

Holmes, Richard. *Shelley: The Pursuit*. New York: Dutton, 1975.

Home, Roderick W. "Electricity and the nervous fluid." *Journal of the History of Biology* 2 (1970): 235–51.

———. "Newton on Electricity and the Aether." In *Contemporary Newtonian Research*. Edited by Zev Bechler. Boston: D. Reidel Publishing Company, 1982: 191–215.

Horlick, Allan S. "Phrenology and the Social Education of Young Men." *History of Education Quarterly* 11, no. 1 (Spring 1971): 23–38.

Hounshell, David A. "Science and Invention." In *Bridge to the Future: A Centennial Celebration of the Brooklyn Bridge*. Edited by Margaret Latimer, Brooke Hindle, and Melvin Kranzberg. New York: New York Academy of Sciences, 1984: 183–92.

Howells, William Dean. *Complete Plays of W. D. Howells*. Edited by Walter J. Meserve. New York: New York University Press, 1960.

———. *Selected Letters of W. D. Howells*, vol. 1 (1852–72), vol. 2 (1873–81), vol. 3 (1882–91), vol. 4 (1892–1901), vol. 5 (1905–11). Edited by George Arms, et al. Boston: Twayne, 1979–83.

Hubert, Philip G., Jr. "The New Talking-Machines." *The Atlantic Monthly* 63, no. 376 (February 1889): 256–61.

Huchinson, William Francis. *The present status of electricity in medicine*. Providence, RI: Providence Press, 1875.

Hudspeth, Robert N., ed. *The Letters of Margaret Fuller*, vol. III, 1842–44. Ithaca: Cornell University Press, 1984.

Hughes, Thomas P. *Networks of Power: Electrification in Western Society, 1880–1930*. Baltimore: Johns Hopkins University Press, 1983.

Hungerford, Edward. "Walt Whitman and His Chart of Bumps." *American Literature* 2, no. 4 (January 1931): 350–84.

Hunt, Alan. "Anxiety and Social Explanation: Some Anxieties about Anxiety." *Journal of Social History* (Spring 1999): 509–28.

Israel, Paul. *Edison: A Life of Invention*. New York: Wiley, 1998.

———. *From Machine Shop to Industrial Laboratory*. Baltimore: Johns Hopkins University Press, 1992.

Jackson, Stanley W. *Melancholia and Depression: From Hippocratic Times to Modern Times*. New Haven: Yale University Press, 1986.

Jacoby, George. "The electrotherapeutic control of currents from central stations." *New York Medical Journal* (3 December 1898 and 10 December 1898).

Jacyna, L. S. "The physiology of mind, the unity of nature, and the moral order in late Victorian thought." *British Journal for the History of Science* 14 (1981): 109–32.

James, Henry. "In the Cage." In *Complete Stories, 1892–1898*. New York: Library of America, 1996: 835–923.

James, William. *Essays in Psychical Research*. Cambridge: Harvard University Press, 1986.

———. *The Varieties of Religious Experience*. Cambridge: Harvard University Press, 1985.

———. *The Will to Believe and other essays in popular philosophy*. New York: Dover, 1956 (1897, 1898).

———. *The Works of William James: Essays, Comments, and Reviews*. Cambridge: Harvard University Press, 1987.

Jastrow, Joseph. "The Problems of 'Psychic Research.'" *Harper's New Monthly Magazine* 79, no. 469 (June 1889): 76–82.

Jennings, Humphrey. *Pandaemonium 1660–1886: The Coming of the Machine As Seen by Contemporary Observers*. Edited by Mary-Lou Jennings and Charles Madge. London: A. Deutsch, 1985.

Johnson, Robert Underwood. *Remembered Yesterdays*. Boston: Little, Brown, 1929.

Jones, Howard Mumford. *The Age of Energy*. New York: Viking, 1971.

Josephson, Matthew. *Edison*. New York: McGraw-Hill, 1959.

Kaplan, Fred. "The Mesmeric Mania: The Early Victorians and Animal Magnetism." *Journal of the History of Ideas* 30 (1974): 691–702.

Katon, W., A. Kleinman, and G. Rosen. "Depression and somatization: A review." *American Journal of Medicine* 72 (1982): 127–34, 241–47.

Kaufman, Martin. *American Medical Education: The Formative Years, 1765–1910*. Westport, CT: Greenwood, 1976.

Keller, Thomas. "Railway Spine Revisited: Traumatic Neurosis or Neurasthenia?" *Journal of the History of Medicine* 50 (1995): 507–24.

Kellogg, John Harvey. "Electrotherapeutics in chronic maladies." *Modern Medicine* (October and November 1904 [n.p.]).

———. *The Graphic Study of Electrical Currents in Relation to Therapeutics*. Battle Creek, MI: Modern Medicine Publishing Co., 1894.

"Kemmler's Terrible Death." *Western Electrician* (10 August 1890): 83–84.

Kennelly, Arthur E. *Biographical Memoir of Thomas Alva Edison*. Washington: National Academy of Sciences, Biographical Memoirs, vol. XV, 1933: 285–304.

———. "Electricity in the Household." *Scribner's Monthly* 7 (1890): 102–15.

Kessler, Joan C., ed. *Demons of the Night*. Chicago: University of Chicago Press, 1995.

Kett, Joseph. *The Formation of the American Medical Profession: The Role of Institutions, 1780–1860*. New Haven: Yale University Press, 1968.

Kiceluk, Stephanie. "The Patient As Sign and Story: Disease Pictures, Life Histories, and the First Psychoanalytic Case History." *Journal of Clinical Psychoanalysis* 1, no. 3 (1992): 333–68.

Kintner, Elvan, ed. *Letters of Robert Browning and Elizabeth Barrett Browning, 1845–46*. Cambridge: Belknap Press of Harvard University Press, 1969.

Kipling, Rudyard. "Wireless." In *Traffics and Discoveries*. New York: Scribner's, 1904 (1902): 239–68.

Kleinman, Arthur. *The Illness Narratives: Suffering, Healing and the Human Condition*. New York: Basic Books, 1988.

———. "Neurasthenia and depression." *Culture, Medicine, and Psychiatry* 6, no. 2 (1982): 117–90.

———. *Rethinking Psychiatry: From Cultural Category to Personal Experience*. New York: Free Press, 1988.

Kleinman, Arthur, and Byron Good, eds. *Culture and Depression*. Berkeley: University of California Press, 1985.

Kline, Ronald. "Construing 'Technology' As 'Applied Science.'" *Isis* 86 (1995): 194–221.

Knight, David. *Humphry Davy: Science & Power*. Oxford: Blackwell, 1992.

———. "The physical sciences and the Romantic movement." *History of Science* 9 (1970): 54–75.

Knight, Nancy. "'The New Light': X Rays and Medical Futurism." In *Imagining Tomorrow: History, Technology, and the American Future*. Edited by Joseph J. Corn. Cambridge: MIT Press, 1986: 10–34.

Kranzberg, Melvin, and William H. Davenport, eds. *Technology and Culture: An Anthology*. New York: New American Library, 1975.

Kuhn, Thomas. *The Essential Tension*. Chicago: University of Chicago Press, 1977.

———. *The Structure of Scientific Revolutions*. Chicago: University of Chicago Press, 1962.

Kusch, Martin. "Recluse, Interlocutor, Interrogator: Natural and Social Order in Turn-of-the-Century Psychological Research Schools." *Isis* 86 (1995): 419–39.

Lake, Henry. "Is Electricity Life?" *Popular Science Monthly* (1873 [n.p.]).

Lawrence, Christopher, and S. L. Shapin, eds. *Science Incarnate: Historical Embodiments of Natural Knowledge*. Chicago: University of Chicago Press, 1998.

Lears, T. Jackson. *No Place of Grace: Antimodernism and the Transformation of American Culture, 1880–1920*. New York: Pantheon, 1981.

Leary, David. "Telling Likely Stories: The Rhetoric of the New Psychology, 1880–1920." *Journal of the History of Behavioral Sciences* 23 (October 1987): 315–31.

Leask, Nigel. "Shelley's 'Magnetic Ladies': Romantic Mesmerism and the Politics of the Body." In *Beyond Romanticism: New Approaches to Texts and Contexts, 1780–1832*. Edited by Stephen Copley and John Whale. London: Routledge, 1992: 53–78.

Leavitt, Judith W., and Ronald L. Numbers, eds. *Sickness and Health in America: Readings in the History of Medicine and Public Health*. Madison: University of Wisconsin Press, 1985.

Lenoir, Timothy. "Models and instruments in the development of electrophysiology, 1845–1912." *Historical Studies in the Physical and Biological Sciences* 17, no. 1 (1986): 1–54.

Levere, Trevor H. *Affinity and Matter: Elements of Chemical Philosophy, 1800–1865*. Oxford: Clarendon, 1971.

———. *Poetry Realized in Nature*. Cambridge: Cambridge University Press, 1981.

Littell, S. "On the Influence of Electrical Fluctuations As a Cause of Disease." *Littell's Living Age* 698 (10 October 1857): 65–77.

Lovering, Joseph. *S. Weir Mitchell*. Boston: Twayne, 1971.

Luckiesh, Matthew. *Artificial Light: Its Influence Upon Civilization*. New York: Century, 1920.

Lungren, Charles M. "Electric and Gas Illumination." *Popular Science Monthly* (September 1882): 577–87.

———. "Electricity at the World's Fair." *Popular Science Monthly* (October 1893): 721–40; (November 1893): 39–54.

Lynn, Michael R. "Enlightenment in the Public Sphere: The Musée de Monsieur and Scientific Culture in Late Eighteenth-Century Paris." *Eighteenth-Century Studies* 32, no. 4 (1999): 463–76.

Mabee, Carleton. *The American Leonardo: A Life of Samuel F. B. Morse*. New York: Knopf, 1943.

MacDonald, Carlos Frederick. *Report of Carlos F. MacDonald on the execution by electricity of William Kemmler alias John Hart*. Albany: Argus, 1890.

Macmillan, M. B. "Beard's Concept of Neurasthenia and Freud's Concept of the Actual Neuroses." *Journal of the History of the Behavioral Sciences* 12, no. 4 (1976): 376–90.

Maines, Rachel P. *The Technology of Orgasm*. Baltimore: Johns Hopkins University Press, 1999.

Makari, George J. "Educated Insane: A Nineteenth-Century Psychiatric Paradigm." *Journal of the History of the Behavioral Sciences* 29 (January 1993): 8–21.

Male, Roy R., Jr. "Hawthorne and the Concept of Sympathy." *PMLA* 68 (1953): 138–49.

———. "Sympathy—a key word in American Romanticism." *The Emerson Society Quarterly* 35 (1964): 19–23.

Marneffe, Daphne de. "Looking and Listening: The Construction of Clinical Knowledge in Charcot and Freud." *Signs* 17, no. 1 (1991): 71–111.

Martin, Thomas Commerford, R. W. Wood, and Elihu Thomson. "Photographing the Unseen: A Symposium on the Roentgen Rays." *The Century* 52, no. 1 (May 1896): 120–31.

Marvin, Carolyn. *When Old Technologies Were New*. New York: Oxford University Press, 1988.

Matlock, Jann. "The Invisible Woman and Her Secrets Unveiled." *Yale Journal of Criticism* 9, no. 2 (1996): 175–221.

Maupassant, Guy de. "The Horla." In *Demons of the Night*. Edited by Joan C. Kessler. Chicago: University of Chicago Press, 1995: 284–308.

McClay, Wilfred M. "Edward Bellamy and the Politics of Meaning." *American Scholar* 64, no. 2 (Spring 1995): 264–71.

McClure, J. B. *Thomas A. Edison and His Inventions*. Chicago: Rhodes & McClure, 1879.

McLaren, Angus. "Phrenology: Medium and Message." *The Journal of Modern History* 46, no. 1 (March 1974): 86–97.

———. "A Prehistory of the Social Sciences: Phrenology in France." *Comparative Studies in Society and History* 23, no. 1 (January 1981): 3–22.

Mendolsohn, Everett. "Physical Models and Physiological Explanation in Nineteenth-Century Biology." *British Journal for the History of Science* 2 (1965): 201–19.

Mertens, Joost. "Shocks and Sparks: The Voltaic Pile As a Demonstration Device." *Isis* 89 (1998): 300–11.

Micale, Mark S. *Approaching Hysteria, Disease and Its Interpretation*. Princeton: Princeton University Press, 1994.

———. "On the 'Disappearance' of Hysteria." *Isis* 84 (1993): 496–526.

Millard, Andre. *Edison and the Business of Invention*. Baltimore: Johns Hopkins University Press, 1990.

Miller, Jonathan. "Going Unconscious." In *Hidden Histories of Science*. Edited by R. B. Silvers. New York: New York Review Press, 1995: 1–35.

Mitchell, Silas Weir. *The Autobiography of a Quack and Other Stories*. New York: Century, 1905.

———. *Fat and Blood: And How to Make Them*. Philadelphia: Lippincott, 1877.

———. *Wear and Tear, or Hints for the Overworked*. Philadelphia: Lippincott, 1871.

Moore, Charles. "Lessons of the Chicago World's Fair. An Interview with the Late Daniel H. Burnham." In *Architecture in America: A Battle of Styles*. Edited by William A. Coles and Henry Hope Reed Jr. New York: Appleton, Crofts, 1961.

Morse, Edward Lind, ed. *Samuel F. B. Morse, His Letters and Journals*, 2 vols. Boston: Houghton Mifflin, 1914.

Morton, William James. "Electricity in Medicine from a Modern Standpoint." *New York Medical Journal* 51 (20 April 1895): 488–92.

Morus, Iwan Rhys. "Currents from the Underworld: Electricity and the Technology of Display in Early Victorian England." *Isis* 84 (1993): 50–69.

———. "The Electric Ariel: Telegraphy and Commercial Culture in Early Victorian England." *Victorian Studies* 39 (1996): 339–78.

———. *Frankenstein's Children*. Princeton: Princeton University Press, 1998.

———. "Marketing the machine: The construction of electro-therapeutics as viable medicine in early Victorian England." *Medical History* 36 (1992): 34–52.

———. "The measure of man: technologizing the Victorian body." *History of Science* 37, part 3, no. 117 (September 1999): 249–82.

Moyer, Albert E. *A Scientist's Voice in American Culture: Simon Newcomb and the Rhetoric of Scientific Method*. Berkeley: University of California Press, 1992.

Neustadter, Roger. "The Deadly Current: The Death Penalty in the Industrial Age." *Journal of American Culture* (Fall 1989): 79–88.

Newcomb, Simon. *His Wisdom, the Defender*. New York: Arno, 1975 (1900).

Newton, Sir Isaac. *Opticks*, 4th ed. New York: McGraw-Hill, 1931 (1730).

Newton, James D. *Uncommon Friends*. New York: Harcourt, Brace, Jovanovich, 1987.

Nitske, W. Robert. *The Life of Wilhelm Conrad Roentgen*. Tucson: University of Arizona Press, 1971.

Nordau, Max. *Degeneration*. New York: Howard Fertig, 1968 (1892, trans. 1895).

———. "A Reply to My Critics." *The Century* 50, no. 4 (August 1895): 546–51.

Noto, Cosimo. *The Ideal City*. New York: Arno, 1971 (1903).

Nye, David. "The Electrified Landscape: A new version of the sublime." *European Contributions to American Studies* 26 (1995): 77–100.

———. *Electrifying America: Social Meanings of a New Technology, 1880–1940*. Cambridge: MIT Press, 1990.

———. *The Invented Self: An Anti-biography from Documents of Thomas Alva Edison*. Odense, Denmark: Odense University Press, 1983.

———. *Narratives and Space: Technology and the Construction of American Culture.* Exeter: University of Exeter Press, 1997.

Olcott, Henry. *Inside the Occult: The True Story of Madame H. P. Blavatsky*. Philadelphia: Running Press, 1975.

Oppenheim, Janet. *"Shattered Nerves": Doctors, Patients, and Depression in Victorian England*. New York: Oxford, 1991.

Ostrom, John Ward, ed. *Letters of Edgar Allan Poe*, 2 vols. Cambridge: Harvard University Press, 1948.

Otis, Laura. "The Metaphoric Circuit: Organic and Technological Communication in the 19th Century." *Journal of the History of Ideas* 63, no. 1 (2002): 105–28.

———. *Networking: Communicating with Bodies and Machines in the Nineteenth Century*. Ann Arbor: University of Michigan Press, 2001.

Palmer, Louis J., Jr. *Encyclopedia of Capital Punishment in the United States*. Jefferson, NC: McFarland, 2001.

Pancaldi, Giuliano. "Electricity and Life: Volta's Path to the Battery." *Historical Studies in Physical and Biological Sciences* 21, no. 1 (1990): 123–60.

Parker, Gail T. *The History of Mind Cure in New England*. Hanover: University Press of New England, 1975.

Parsons, Gail Pat. "Equal Treatment for All: American Medical Remedies for Male Sexual Problems: 1850–1900." *Journal of the History of Medicine* (January 1977): 55–71.

Paser, Harold Clarence. *The Electrical Manufacturers, 1875–1900*. New York: Arno, 1972.

Peck, Bradford. *The World a Department Store*. Lewiston, ME: Bradford Peck, 1900.

Penrose, James F. "Inventing Electrocution." *The Current Commentary* (May–September 1995), n.p.

Pera, Marcello. *The Ambiguous Frog: The Galvanism-Volta Controversy on Animal Electricity*. Translated by Jonathan Mandelbaum. Princeton: Princeton University Press, 1992.

Perkins, George. "The Physician of the Future." *Popular Science Monthly* (September 1882): 637–45.

Perry, J. "The Future Development of Electrical Appliances." *Journal of the Society of the Arts* 29 (1880–81): 457.

Peterson, M. Jeanne. *The Medical Profession in Mid-Victorian London*. Berkeley: University of California Press, 1978.

Pfeiffer, Carl J. *The Art and Practice of Western Medicine in the Early Nineteenth Century*. Jefferson, NC: McFarland, 1985.

Pichanick, Valerie K. *Harriet Martineau, the Woman and Her Work*. Ann Arbor: University of Michigan Press, 1980.

Pinero, José M. Lopez. *Historical Origins of the Concept of Neurosis*. Cambridge: Cambridge University Press, 1983.

Piper, Herbert Walter. *The Active Universe*. London: Athlone, 1962.

Platt, Harold. *The Electric City: Energy and the Growth of the Chicago Area, 1880–1930*. Chicago: University of Chicago Press, 1991.

Poe, Edgar Allan. *The Science Fiction of Edgar Allan Poe*. Edited by Harold Beaver. New York: Penguin, 1976.

Poirier, Suzanne. "The Weir Mitchell Rest Cure: Doctors and Patients." *Women's Studies* 10, no. 1 (1983): 15–40.

Porter, Roy. *Madness, A Brief History*. Oxford: Oxford University Press, 2002.

Prescott, George B. *The Speaking Telephone, Talking Phonograph and Other Novelties*. New York: Appleton, 1878.

Priestley, Joseph. *The History and Present State of Electricity*, 3d ed., 2 vols. New York: Johnson Reprint Corporation, 1966 [1755].

Prince, Morton. "The True Position of Electricity As a Therapeutic Agent in Medicine." *Boston Medical and Surgical Journal* 123, no. 14 (2 October 1890): 313–19.

"Professor Brackett Opposed to Executions by Electricity." *Electrical Review* 28 (December 1888).

Pursell, Carroll W. *Readings in Technology and American Life*. New York: Oxford University Press, 1969.

Rabinbach, Anson. *The Human Motor: Energy, Fatigue, and the Origins of Modernity*. New York: Basic Books, 1990.

Radden, Jennifer, ed. *The Nature of Melancholy*. New York: Oxford University Press, 2000.

Rauch, Alan. *Useful Knowledge: The Victorians, Morality, and the March of Intellect*. Durham, NC: Duke, 2001.

Rein, David M. *S. Weir Mitchell As a Psychiatric Novelist*. New York: International Universities Press, 1952.

Reiser, Stanley Joel. *Medicine and the Reign of Technology*. Cambridge: Cambridge University Press, 1978.

Reynolds, Edward. "On the Value of Electricity in Minor Gynecology." *Boston Medical and Surgical Journal* 126, no. 16 (21 April 1892): 381–85.

Riegel, Robert E. "The Introduction of Phrenology to the United States." *American Historical Review* 39, no. 1 (October 1933): 73–78.

Ritterbush, Phillip C. "Electricity: The Soul of the Universe." In *Overtures to Biology*. New Haven: Yale University Press, 1964.

Rockwell, Alphonso David. *Rambling Recollections: An Autobiography*. New York: Paul B. Holbin, 1920.

Rodin, Alvin, and Jack Key. *Medical Casebook of Doctor Arthur Conan Doyle*. Malabar, FL: Robert E. Krieger, 1984.

Ronnell, Avital. *The Telephone Book: Technology, Schizophrenia, Electric Speech*. Lincoln: University of Nebraska Press, 1989.

Rose, Mark H. *Cities of Light and Heat: Domesticating Gas and Electricity in American Homes*. Philadelphia: University of Pennsylvania Press, 1995.

Rosenberg, Charles E. "Body and Mind in Nineteenth-Century Medicine: Some Clinical Origins of the Neurosis Construct." *Bulletin of the History of Medicine* 63 (1989): 185–97.

———. *Explaining Epidemics and Other Studies in the History of Medicine*. Cambridge: Cambridge University Press, 1992.

———. *No Other Gods: On Science and American Social Thought*. Baltimore: Johns Hopkins University Press, 1976.

———. "Pathologies of Progress: The Idea of Civilization As Risk." *Bulletin of the History of Medicine* 72, no. 4 (1998): 714–30.

———. "The Place of G. M. Beard in Nineteenth-Century Psychiatry." *Bulletin of the History of Medicine* 36 (1962): 245–59.

———. *The Trial of the Assassin Guiteau*. Chicago: University of Chicago Press, 1968.

Rosenberg, Charles E., and Janet Golden, eds. *Framing Disease: Studies in Cultural History*. New Brunswick: Rutgers University Press, 1992.

Rosner, Lisa. "The Professional Context of Electrotherapeutics." *Journal of the History of Medicine* 43 (1988): 64–82.

Roth, Nancy. "The great patent medicine era." *Medical Instrumentation* 11, no. 5 (September–October 1977): 302–3.

———. "Tracking by telephone: locating the bullet in President Garfield, 1881." *Medical Instrumentation* 15, no. 3 (May–June 1981): 190.

Rothfield, Lawrence. *Vital Signs: Medical Realism in Nineteenth Century Fiction*. Princeton: Princeton University Press, 1992.

Rothstein, William G. *American Medical Schools and the Practice of Medicine*. New York: Oxford University Press, 1987.

———. *American Physicians in the Nineteenth Century: From Sects to Science*. Baltimore: Johns Hopkins University Press, 1972.

Rousseau, G. S. "Discourses of the Nerve." In *Literature and Science As Modes of Expression*. Edited by Frederick Amrine. Dordrecht, Netherlands: Kluwer Academic, 1989.

Rowbottom, Margaret, and Charles Susskind. *Electricity and Medicine: History of Their Interaction*. San Francisco: San Francisco Press, 1984.

Ruddick, Nicholas. "Life and Death by Electricity in 1890: The Transfiguration of William Kemmler." *Journal of American Culture* 21, no. 4 (1998): 79–87.

Rydell, Robert W. *All the World's a Fair*. Chicago: University of Chicago Press, 1984.

Sadoff, Dianne F. *Sciences of the Flesh: Representing Body and Subject in Psychoanalysis*. Stanford: Stanford University Press, 1998.

Salomon, Jean-Jacques. "Public Reactions to Science and Technology: The Wizard Faces Social Judgment." In *Science Technology and the Human Prospect*. Edited by C. Starr and P. Ritterbush. New York: Pergamon, 1980: 77–93.

Sappol, Michael. *A Traffic of Dead Bodies*. Princeton: Princeton University Press, 2002.

Sarton, George. "The Discovery of X-Rays." *Isis* 26, no. 2 (March 1937): 349–69.

Schaffer, Simon. "The Consuming Flame: electrical showmen and Tory mystics in the world of goods." In *Consumption and the World of Goods*. Edited by John Brewer and Roy Porter. London: Routledge, 1992: 489–526.

———. "Natural Philosophy and Public Spectacle in the Eighteenth Century." *History of Science* 21, no. 1 (1983): 1–43.

———. "Self evidence." *Critical Inquiry* 18 (Winter 1992): 327–62.

Schivelbusch, Wolfgang. *Disenchanted Night*. Translated by Angela Davies. Berkeley: University of California Press, 1988.

———. *The Railway Journey: The Industrialization of Time and Space in the Nineteenth Century*. Berkeley: University of California Press, 1986.

Schlereth, Thomas J. *Victorian America: Transformations in Everyday Life*. New York: HarperCollins, 1991.

Schroeder, Fred E. H. "More 'Small Things Forgotten': Domestic Electrical Plugs and Appliances, 1881–1931." *Technology and Culture* 27, no. 3 (1986): 525–43.

Scull, Andrew D. *Madhouses, Mad-Doctors and Madmen: The Social History of Psychiatry in the Victorian Era*. Philadelphia: University of Pennsylvania Press, 1981.

Secord, James A. "Extraordinary Experiment: electricity and the creation of life in Victorian England." In *The Uses of Experiment*. Edited by David Gooding, Trevor Pinch, and Simon Schaffer. Cambridge: Cambridge University Press, 1989: 337–83.

Seltzer, Mark. *Bodies and Machines*. New York: Routledge, 1992.

Shiers, George, ed. *The Electric Telegraph: An Historical Anthology*. New York: Arno, 1977.

Shorter, Edward. *From Paralysis to Fatigue: A History of Psychosomatic Illness in the Modern Era*. New York: Free Press, 1992.

Shortt, S. E. D. "Physicians and Psychics: The Anglo-American Medical Response to Spiritualism, 1870–1890." *Journal of the History of Medicine and Allied Sciences* 39 (1984): 339–55.

Shyrock, Richard H. *The Development of Modern Medicine*. Philadelphia: University of Pennsylvania Press, 1936.

Sicherman, Barbara. "The Paradox of Prudence: Mental Health in the Gilded Age." *Journal of American History* 62, no. 4 (March 1976): 890–912.

———. "The Uses of Diagnosis: Doctors, Patients, and Neurasthenia." *Journal of the History of Medicine* 32, no. 1 (1977): 33–54.

Sicilia, David B. "Selling Power: Market and Monopoly at Boston Edison, 1886–1926." *Business and Economic History* 20 (1991): 27–31.

Siefert, Marsha. "Aesthetics, Technology, and the Capitalization of Culture: How the Talking Machine Became a Musical Instrument." *Science in Context* 8, no. 2 (1995): 417–49.

Sinclair, Bruce. "Technology on Its Toes: Late Victorian Ballets, Pageants, and Industrial Exhibitions." In Stephen H. Cuttliffe and Robert C. Post. *In Context: History and the History of Technology*. Bethlehem: Lehigh University Press, 1989: 71–87.

Smith, Henry Nash, ed. *Popular Culture and Industrialism, 1865–1890.* New York: New York University Press, 1967.

Smith, Henry Nash, and William M. Gibson, eds. *Mark Twain–Howells Letters, 1872–1910*, 2 vols. Cambridge: Harvard University Press, 1960.

Stainbrook, Edward. "The Uses of Electricity in Psychiatric Medicine during the Nineteenth Century." *Bulletin of the History of Medicine* 22 (1948): 164–75.

Standage, Tom. *The Victorian Internet.* New York: Walker, 1998.

Starr, M. Allen. "Electricity in Relation to the Human Body." *Scribner's Monthly* 6 (1889): 589–99.

Starr, Paul. *The Social Transformation of American Medicine.* New York: Basic Books, 1982.

Staudenmaier, John M. *Technology's Storytellers: Reweaving the Human Fabric.* Cambridge: MIT Press, 1985.

Stauffer, Robert C. "Persistent Errors Regarding Øersted's Discovery of Electromagnetism." *Isis* 44, no. 4 (December 1953): 307–10.

———. "Speculation and Experiment in the Background of Øersted's Discovery of Electromagnetism." *Isis* 48 (1957): 33–50.

Stern, Madeleine B. "Mark Twain Had His Head Examined." *American Literature* 41, no. 2 (May 1969): 207–18.

Stevenson, William G. "Physiological Significance of Vital Force." *Popular Science Monthly* 24 (April 1884): 760–73.

Stookey, Byron. "A Lost Neurological Society with Great Expectations." *Journal of the History of Medicine* 16 (July 1961): 280–93.

Strindberg, August. *By the Open Sea.* Translated by Ellie Schleussner. New York: Haskell House, 1973 [1890].

Strouse, Jean. *Morgan: American Financier.* New York: Random House, 1999.

Sunstein, Emily. *Mary Shelley, Romance and Reality.* Boston: Little, Brown, 1989.

Sutton, Geoffrey. "Electrical Medicine and Mesmerism." *Isis* 72 (1981): 375–92.

Tatar, Maria M. *Spellbound: Studies on Mesmerism and Literature.* Princeton: Princeton University Press, 1978.

Taylor, Eugene. "The electrified hand—psychotherapeutic implications." *Medical Instrumentation* 17, no. 4 (July–August 1983): 281–82.

Teich, Mikulas, and Roy Porter, eds. *Fin-de-Siècle and Its Legacy.* Cambridge: Cambridge University Press, 1990.

"The Telegraph." *Harper's New Monthly Magazine* 47, no. 279 (April 1873): 332–60.

Temkin, Owsei. "Basic Science, Medicine, and the Romantic Era." *Bulletin of the History of Medicine* 37 (1963): 97–129.

Thompson, Robert Luther. *Wiring a Continent: The History of the Telegraph Industry in the United States, 1832–1866.* Princeton: Princeton University Press, 1947.

Tichi, Cecelia. *Shifting Gears: Technology, Literature, Culture in Modernist America.* Chapel Hill: University of North Carolina Press, 1987.

Tocco, Peter. "The Night They Turned the Lights On in Wabash." *Indiana Magazine of History* 95, no. 4 (December 1999): 350–63.

Townsend, Horace. "Edison, His Work, and His Work-Shop." *Cosmopolitan* (April 1889): 598–607.

Trachtenberg, Alan. *The Incorporation of America*. New York: Hill and Wang, 1982.

Trowbridge, John. *The Electrical Boy, or, The Career of Greatman and Greatthings*. Boston: Roberts Brothers, 1891.

Twain, Mark. *The American Claimant*. New York: Oxford University Press, 1996 (1892).

———. *Mark Twain's Notebooks and Journals*, vol III. Edited by Robert Pack Browning, Michael Frank, and Lin Salamo. Berkeley: University of California Press, 1979.

Ueyama, Takahiro. "Capital, Profession, and Medical Technology. The Electrotherapeutic Institutes and the Royal College of Physicians." *Medical History* 41 (April 1997): 150–81.

Vanable, Joseph W., Jr. "A History of Bioelectricity in Development and Regeneration." In *A History of Regeneration*. Edited by C. Dinsmore. Cambridge: Cambridge University Press, 1991.

Van Deusen, Edward Holmes. "Observations on a Form of Nervous Prostration (Neurasthenia) Culminating in Insanity." *American Journal of Insanity* 25 (1868–69): 445–61.

Verschuur, Gerrit L. *Hidden Attraction: The History and Mystery of Magnetism*. Oxford: Oxford University Press, 1993.

Vila, Bryan, and Cynthia Morris, eds. *Capital Punishment in the United States: A Documentary History*. Westport, CT: Greenwood, 1997.

Villiers de L'Isle-Adam, Auguste de. *Tomorrow's Eve*. Translated by R. M. Adams. Urbana: University of Illinois Press, 1982 (1886).

Vincenti, Walter G. "The technical shaping of technology: Real-world constraints and technical logic in Edison's electric lighting system." *Social Studies of Science* 25 (1995): 553–74.

Vogel, Morris J., and Charles E. Rosenberg, eds. *The Therapeutic Revolution*. Philadelphia: University of Pennsylvania Press, 1979.

Vrettos, Athena. *Somatic Fictions: Imagining Illness in Victorian Culture*. Stanford, CT: Stanford University Press, 1995.

Wachhorst, Wyn. *Thomas Alva Edison: An American Myth*. Cambridge: MIT Press, 1981.

Waddington, Ivan. *The Medical Profession in the Industrial Revolution*. Dublin: Gill & Macmillan, 1984.

Wakefield, Edward. "Nervousness: The National Disease of America." *McClure's* 2 (1893–94): 302–7.

Walker, Sydney F. *Electricity in Our Homes and Workshops*. London: Whittaker, 1889.

Walter, Dave, ed. *Today Then: America's Best Minds Look 100 Years into the Future on the Occasion of the 1893 World's Columbian Exposition*. Helena, MT: American & World Geographic Publishing, 1992.

Warner, John Harley. *The Therapeutic Perspective: Medical Practice, Knowledge, and Identity in America, 1820–1885*. Princeton: Princeton University Press, 1997.

Washburn, George R. "The Sedative Action of Electricity." *Boston Medical and Surgical Journal* (31 October 1889): 430–32.

Webb, R. K. *Harriet Martineau: A Radical Victorian*. New York: Columbia University Press, 1960.

Weiner, Philip P. "G. M. Beard and Freud on 'American Nervousness.'" *Journal of the History of Ideas* 17 (1956): 269–74.

Westinghouse, George. "A Reply to Mr. Edison." *North American Review* (December 1889): 653–64.

Wheatley, Vera. *The Life and Work of Harriet Martineau*. London: Secker & Warburg, 1957.

Whittaker, Sir Edmund. *A History of the Theories of Aether and Electricity, Part I: The Classical Theories*. New York: American Institute of Physics, 1987 (1951), n.p.

Wiggin, Kate Douglas. "Philippa's Nervous Prostration; a Study in Nobleness." *Scribner's Monthly* 39 (1906): 1–14.

Williams, Henry Smith. "Science at the Beginning of the Century." *Harper's New Monthly Magazine* 94, no. 560 (January 1897): 217–18.

Williams, J. P. "Psychical research and psychiatry in late Victorian Britain: trance as ecstasy or trance as insanity." In *The Anatomy of Madness*, vol. 1. Edited by W. F. Bynum, Roy Porter, and Michael Shepherd. London: Tavistock, 1985: 233–54.

Williams, L. Pearce. *Michael Faraday*. New York: Simon & Schuster, 1971.

Williams, W. Mattieu. "Electromania." *Popular Science Monthly* (September 1882): 650–55.

Winter, Alison. "Mesmerism and Popular Culture in Early Victorian England." *History of Science* 32 (1994): 317–43.

———. *Mesmerized: Powers of Mind in Victorian Britain*. Chicago: University of Chicago Press, 1998.

Withington, Vere. "An Electrical Study; a story." *Overland Monthly* 20 (October 1892): 417–29.

Wolkomir, Richard, and Joyce Wolkomir. "Mr. Edison Takes a Holiday." *Smithsonian* 30, no. 9 (December 1999): 136–51.

Wood, Ann Douglas. "Women's Complaints and Their Treatment in Nineteenth-Century America." In *Clio's Consciousness Raised: New Perspectives on the History of Women*. Edited by Mary S. Hartman and Lois Banner. New York: Harpers, 1974.

Wood, Gaby. *Edison's Eve*. New York: Knopf, 2002.

Wood, Henry. *The New Thought Simplified: how to gain harmony and health*. Boston: Lee & Shepard, 1903.

Woodward, John, and David Richards, eds. *Health Care and Popular Medicine in Nineteenth-Century England*. New York: Holmes & Meier, 1977.

Woodward, William R., and Mitchell G. Ash, eds. *The Problematic Science: Psychology in Nineteenth-Century Thought*. New York: Praeger, 1982.

Wrobel, Arthur, ed. *Pseudoscience and Society in Nineteenth-Century America*. Louisville: University of Kentucky Press, 1987.

Wynne, Brian. "Physics and Psychics: Science, Symbolic Action and Social Control in Late Victorian England." In *Natural Order: Historical Studies of Scientific Culture*. Edited by Barry Barnes and Steven Shapin. Beverly Hills and London: Sage, 1979: 167–86.

Index